Design of Analog Circuits

Dr. A.V.N.Tilak

KHANNA BOOK PUBLISHING CO. (P) LTD.

Publisher of Engineering and Computer Books

4C/4344, Ansari Road, Darya Ganj, New Delhi-110002

Phone : 011-23244447-48 **Mobile:** +91-9910909320

E-mail : contact@khannabooks.com

Website : www.khannabooks.com

Dear Readers,

To prevent the piracy, this book is secured with HIGH SECURITY HOLOGRAM on the front title cover. In case you don't find the hologram on the front cover title, please write us to at contact@khannabooks.com or whatsapp us at +91-99109 09320 and avail special gift voucher for yourself.

Specimen of Hologram on front Cover title:

Moreover, there is a SPECIAL DISCOUNT COUPON for you with EVERY HOLOGRAM.

How to avail this SPECIAL DISCOUNT:

Step 1: Scratch the hologram
Step 2: Under the scratch area, your "coupon code" is available
Step 3: Logon to www.khannabooks.com
Step 4: Use your "coupon code" in the shopping cart and get your copy at a special discount
Step 5: Enjoy your reading!

Price: ₹ 498.00

ISBN: 978-93-91505-00-4

Design of Analog Circuits *by*
Dr. A.V.N. Tilak

First Edition: 2022
Reprint: 2024, 2026

To view complete list of books, Please scan the QR Code:

KPH

Published by:
Khanna Book Publishing Co. (P) Ltd.
CIN: U22110DL1998PTC095547

Visit us at: **www.khannabooks.com**
Write us at: **contact@khannabooks.com**

Printed in Bharat.

Preface

Electronics, the most exciting area of technology nowadays, is playing a vital role in our daily life. It is becoming difficult to carry out our works without using electronic systems. The present generation uses robots and artificial intelligence to do human work with ease and efficiency. There is no doubt that many functions that were traditionally the domain of analog electronics are at present implemented in digital form. However, the physical world is inherently analog, indicating that there will always be a need for analog circuitry to condition physical signals such as those associated with transducers, as well as to convert information from analog to digital for processing, and from digital back to analog for reuse in the physical world. Moreover, new applications continue to emerge, where considerations of speed and power make it more advantageous to use analog front ends. Indeed, many applications today are best addressed by mixed-mode ICs and systems, which rely on analog circuitry to interface with the physical world.

Even though the analog circuitry may constitute only a small portion of the total chip area, it is often the most challenging part to design as well as the limiting factor on the performance of the entire system. In this respect, it is usually the analog designer who is called to devise ingenious solutions to the task of realizing analog functions in decidedly digital technologies. Analog sensors are used in smart phone to interface with its advanced processor. The internet of things makes use of an array of sensors and ICs many of which are interconnected using analog circuits that enables the wide collection and processing of large amounts of real-time data. There is an increasing demand for analog modules in these and other systems that carry out a variety of functions. Even purely digital circuits, when pushed to their operational limits, exhibit analog behavior. Consequently, a solid grasp of analog design principles and techniques is a valuable asset in the design of any IC, not just purely digital or purely analog ICs. Therefore, there is a critical need to develop skills for the design of analog electronic circuits. In light of these, the need for competent analog designers will continue to remain very strong.

There are several books available in the market that deal extensively with analog electronic circuits, but most tend to give theory, explain the circuit behavior, and analysis. They generally do not enable the reader to design complete practical circuits. This text addresses this shortcoming by treating the subject comprehensively. It first provides the design specifications, the theory and operation of electronic circuits using diode, metal oxide semiconductor field effect transistor (MOSFET), bipolar junction transistor (BJT), operational amplifier, 555 timer, and voltage regulator ICs. It then presents the equations to be made use of in the design of circuits, and the design of practical circuits of varying levels of complexity including regulated power supplies, clippers, clampers, voltage amplifiers, feedback amplifiers, power amplifiers, oscillators, filters, circuits for arithmetic operations, and waveform generators. After going through the book, the reader will understand the operation and be able to confidently design a broad range of functioning analog circuits and systems.

Design is the heart of engineering and it remains a difficult issue in educating engineers. The use of the well-defined problem-solving methodology presented in this text can significantly enhance the students' ability to understand issues related to design. This book enables the reader to develop a deep understanding of the operation, and to design a range of working circuits. Many examples help the reader to quickly become familiar with key design parameters and design methodology for each class of circuits. With the proliferation of electronic systems today, the need for electronic design engineers is greater than before, and it can be believed that this book fills an important niche in the world of analog electronics. It is intended primarily as an undergraduate text for university and college students enrolled in electrical, electronic, and computer engineering programs. It should also prove useful as a reference book for practicing engineers who need to design practical systems to meet research or real-world needs.

The book is organized into four chapters. In chapter 1, the design of diode circuits is presented. Chapter 2 covers MOSFET circuits design. The design of BJT circuits are given in chapter 3. Design of circuits using operational amplifiers and linear ICs are discussed in chapter 4. Finally, the datasheets of devices and ICs used in the design are given in the appendices at the end. The design of all of the circuits covered are important in today's electronics and learning about these should be an enjoyable and rewarding experience. I hope this book provides that kind of experience for its readers.

Dr. A.V.N. Tilak

Contents

1

CHAPTER

DIODE CIRCUITS

DIODE CIRCUITS

Diode is a two-terminal device and is fabricated by joining *P*- and *N*-type semiconductor materials, forming a PN junction. The flow of charge carriers across the junction develops depletion potential. The diode will not conduct until the anode is made more positive than the cathode by a forward voltage sufficient to overcome the depletion potential. In the forward bias conditions, the current through the diode increases in an exponential manner. When the cathode is more positive than the anode, the diode is reverse biased and will not conduct current. Under these conditions, a very small reverse leakage current flows through the diode.

Diode is a simplest and most fundamental non-linear circuit element. Many types of diodes are available and are used for a wide range of applications. Diodes are useful as rectifiers to convert AC line voltage into DC, in clipping and clamping circuits to shape and modify AC signals, in generation of various types of signals, and in AM detectors. Zener diodes are used to regulate output DC voltages and as a voltage reference in electronic circuits. Light emitting diodes (LEDs) are used in opto-electronic applications, data display boards, traffic light indicators, medical applications, remote controls, etc. Photodiodes are used as receivers for infrared remote control devices, in opto-electronic applications, medical devices, compact disc players, and smoke detectors.

In this chapter, design of, rectifiers with capacitor filter, clippers, clampers, voltage doubler, function generator for voltage transfer characteristics and *V-I* characteristics, waveform generator, AM demodulator, Zener voltage regulator, optocoupler using LED and photodiode, for given specifications, is given.

1.1 DC POWER SUPPLY USING CENTER TAPPED FULL-WAVE RECTIFIER WITH CAPACITOR FILTER

Design Specifications

(*i*) DC output voltage = 9 V at 500 mA (*ii*) Ripple = 5% maximum.

Theory

Fig. 1.1(*a*) shows the full-wave rectifier with capacitor filter. It utilizes both halves of the input sinusoid and provides a unipolar output by inverting the negative halves of the sine wave. The transformer secondary winding is center tapped to provide two equal voltages V_s across the two halves of the secondary winding with the polarities indicated as shown in Fig. 1.1(*b*).

When the input line voltage feeding the primary is positive, both of the signals labelled V_s will be positive. In this case D_1 will conduct and D_2 will be reverse biased. The current through D_1 will flow through C and R_L and back to the center tap of the secondary. Now, during the negative half cycle of the a.c. line voltage, both of the voltages labelled V_S will be negative. Thus, D_1 will be cut OFF while D_2 conducts. The current conducted by D_2 will flow through C and R_L and back to the center tap. The current through R_L always flows in the same direction, and thus V_o will be unipolar, as indicated in Fig. 1.1(*c*).

For a sinusoidal input, the capacitor charges to the peak of the input V_m. Then the diode cuts OFF, and the capacitor discharges through the load resistance R_L. The capacitor discharge will continue for almost the entire cycle, until the time at which V_S exceeds the capacitor voltage. Then the diode turns ON again and charges the capacitor up to the peak of V_S and the process repeats itself. To keep the output voltage from decreasing too much during capacitor discharge, the value of C is chosen so that the time constant CR_L is much greater than the discharge interval. The output voltage with capacitor filter is shown in Fig. 1.1(*d*). Here T_1 is the capacitor charging time, T_2 is the capacitor discharging time, and V_{rpp} is the peak-to-peak value of ripple voltage.

Circuit Diagram

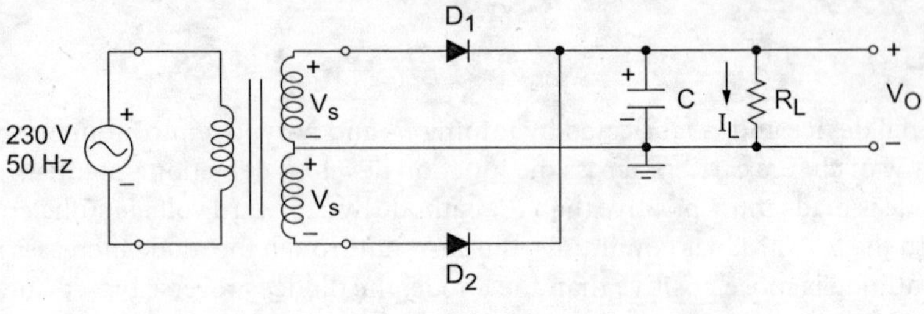

(*a*)

Waveforms

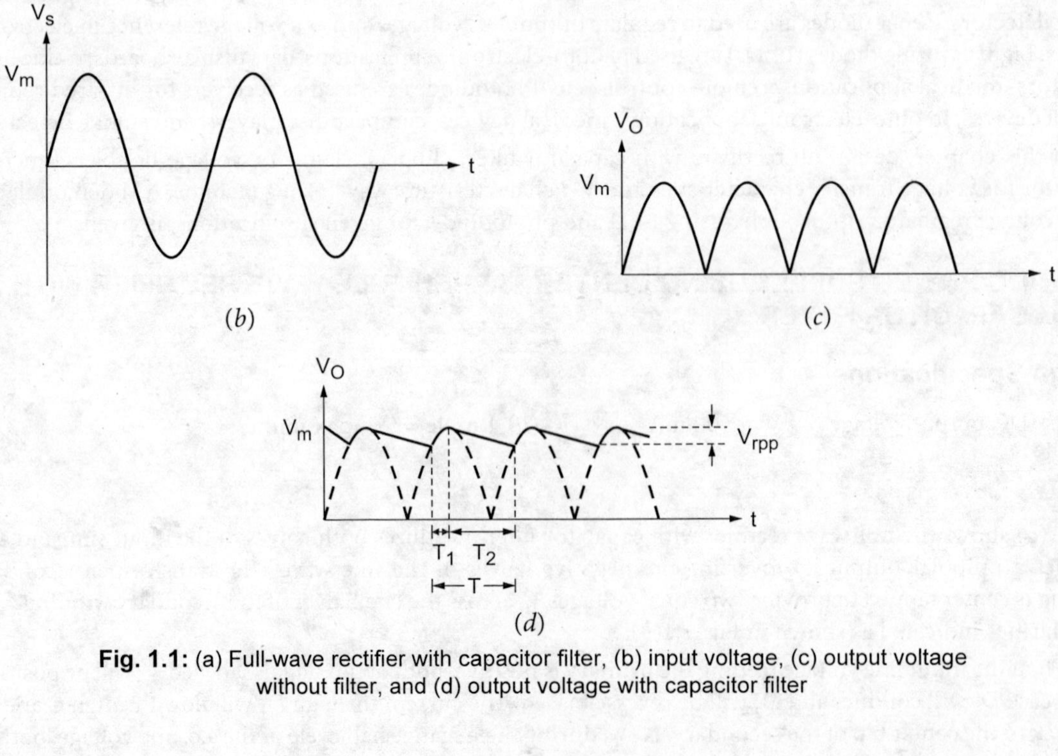

(*b*) (*c*) (*d*)

Fig. 1.1: (a) Full-wave rectifier with capacitor filter, (b) input voltage, (c) output voltage without filter, and (d) output voltage with capacitor filter

Design Equations

Load resistance $R_L = \frac{V_O}{I_L}$, V_O is the output DC voltage and I_L is the output or load current.

The transformer secondary voltage, $V_s = V_m \sin \omega t$

The peak or maximum value of transformer secondary voltage $V_m = V_O + V_{rpp}/2 + V_D$ where V_{rpp} is the peak-to-peak ripple voltage and V_D is the diode drop under forward bias conditions.

$V_{rpp} = \frac{V_m - V_D}{2fCR_L}$, f is the input frequency = 50 Hz and C is the filter capacitor.

Hence,
$$C = \frac{V_m - V_D}{2V_{rrp} fR_L}$$

The peak diode current is given by

$$i_D\,(\text{peak}) = \left(\frac{V_m}{R_L}\right)\left(1+\pi\sqrt{\frac{2V_m}{V_{rpp}}}\right)$$

The average diode current is given as

$$i_D\,(\text{av}) = \left[\left(\frac{1}{\pi}\right)\sqrt{\frac{2V_{rpp}}{V_m}}\right]\left(\frac{V_m}{R_L}\right)\left[1+\left(\frac{\pi}{2}\right)\sqrt{\frac{2V_m}{V_{rpp}}}\right]$$

The peak inverse voltage that each diode must sustain is

$$PIV = 2V_m$$

Design Steps

Load resistance, $R_L = 18\ \Omega$

Power rating of R_L, $P_{RL} = I_L^2 R_L = 4.5$ W

Chose, $R_L = 18\ \Omega$, 5 W

For a 5 percent ripple, the ripple voltage $V_{rpp} = (0.05)V_O = 0.45$ V

Assuming silicon diodes with a forward voltage drop of 0.7 V, the peak value of v_S is

$$V_m = 9.925 \text{ V}$$

For a sinusoidal signal, this gives an rms voltage of $V_{Srms} = 7.02$ V

Chose a center tapped transformer of 230V/(7.5-0-7.5)V, 500 mA.

The transformer turns ratio, $\frac{N_P}{N_S} = 30.67 \approx 31$

The required filter capacitor $C = 0.01139$ F = 11390 μF

Choose $C = 12000$ μF, 16 V

The peak diode current = 12.06 A

The average diode current = 0.6 A

Finally, for the transformer secondary voltage of 7.5 V, the peak inverse voltage that each diode must sustain is

$$PIV = 2V_m = 21.21 \text{ V}$$

The diode datasheets specify 1N4001 diode can meet these requirements. Chose 2 numbers of 1N4001 diodes.

1.2 DC POWER SUPPLY USING BRIDGE RECTIFIER WITH CAPACITOR FILTER

Design Specifications

(*i*) DC output voltage = 9 V at 500 mA (*ii*) Ripple = 5% maximum.

Theory

Figure 1.2(*a*) shows the bridge rectifier with capacitor filter. It utilizes both halves of the input sinusoid and provides a unipolar output by inverting the negative halves of the sine wave. When the input line voltage feeding the primary is positive, the signal labelled V_s will be positive as shown in Fig. 1.2(*b*). In this case D_2 and D_3 will conduct and D_1 and D_4 will be reverse biased. The current through D_2 will flow through C, R_L, D_3 and back to the secondary of transformer. Now, during the negative half cycle of the ac line voltage, the voltage labelled V_S will be negative. Thus, D_2 and D_3 will be cut OFF while D_1 and D_4 conducts. The current conducted by D_4 will flow through C, R_L, D_1 and back to the transformer secondary. The current through R_L always flows in the same direction, and thus V_o will be unipolar, as indicated in Fig. 1.2(*c*).

For a sinusoidal input, the capacitor charges to the peak of the input V_m. Then the diode cuts OFF, and the capacitor discharges through the load resistance R_L. The capacitor discharge will continue for almost the entire cycle, until the time at which V_S exceeds the capacitor voltage. Then two diodes will turn ON again and charges the capacitor up to the peak of V_S and the process repeats itself. To keep the output voltage from decreasing too much during capacitor discharge, the value of C is chosen so that the time constant CR_L is much greater than the discharge interval. The output voltage with capacitor filter is shown in Fig. 1.2(*d*). Here T_1 is the capacitor charging time, T_2 is the capacitor discharging time, and V_{rpp} is the peak-to-peak value of ripple voltage.

Circuit Diagram

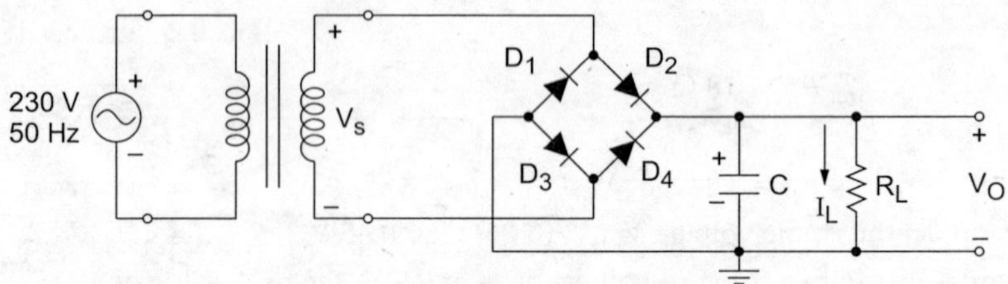

(*a*)

Waveforms

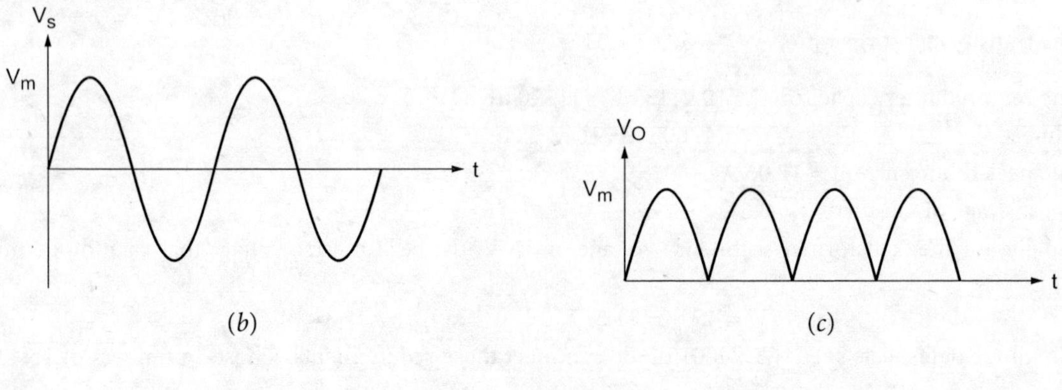

(*b*) (*c*)

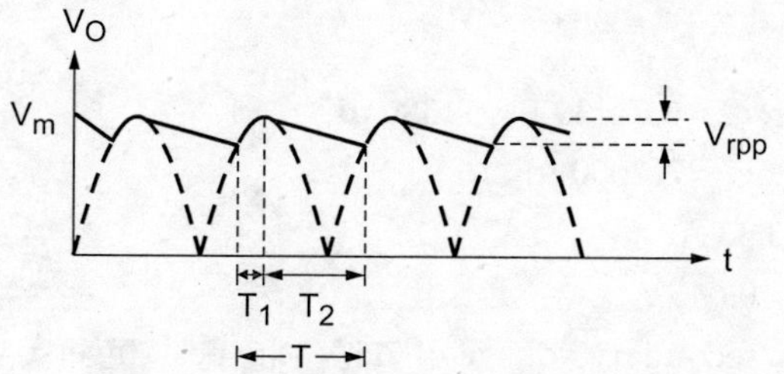

(*d*)

Fig. 1.2: (a) Bridge rectifier with capacitor filter, (b) input voltage, (c) output voltage without filter, and (d) output voltage with capacitor filter

Design Equations

Load resistance $R_L = \dfrac{V_o}{I_L}$, V_o is the output DC voltage and I_L is the output or load current.

The transformer secondary voltage, $V_s = V_m \sin \omega t$

The peak value of transformer secondary voltage, $V_m = V_o + V_{rpp}/2 + 2V_D$

where V_{rpp} is the peak-to-peak ripple voltage and V_D is the diode drop under forward bias conditions.

$V_{rpp} = \dfrac{V_m - 2V_D}{2fCR_L}$, f is the input frequency = 50 Hz and C is the filter capacitor.

Hence, $$C = \frac{V_m - 2V_D}{2V_{rpp} fR_L}$$

The peak diode current is given by

$$i_D\,(\text{peak}) = \left(\frac{V_m}{R_L}\right)\left(1+\pi\sqrt{\frac{2V_m}{V_{rpp}}}\right)$$

The average diode current is given as

$$i_D\,(\text{peak}) = \left[\left(\frac{1}{\pi}\right)\sqrt{\frac{2V_{rpp}}{V_m}}\right]\left(\frac{V_m}{R_L}\right)\left[1+\left(\frac{\pi}{2}\right)\sqrt{\frac{2V_m}{V_{rpp}}}\right]$$

The peak inverse voltage that each diode must sustain is

$$PIV = V_m$$

Design Steps

Load resistance, $R_L = 18\ \Omega$

Power rating, $P_{RL} = I_L^2 R_L = 4.5$ W

Chose, $R_L = 18\ \Omega$, 5 W

For a 5 percent ripple, the ripple voltage is $V_{rpp} = 0.45$ V

Assuming silicon diodes with a forward voltage drop V_D of 0.7 V, the peak value of V_S is

$$V_m = 10.625\ \text{V}$$

For a sinusoidal signal, this produces an rms voltage of $V_{Srms} = 7.513$ V

Chose a transformer of 230V/7.5V, 500 mA.

The transformer turns ratio is $\frac{N_P}{N_S} = 30.67 \approx 31$

The required filter capacitor, $C = 0.01139\text{ F} = 11390\ \mu\text{F}$

Choose $C = 12000\ \mu\text{F, 16 V}$

The peak diode current = 13.33 A

The average diode current = 0.645 A

Finally, for the transformer secondary voltage of 7.5 V, the peak inverse voltage that each diode must sustain is

$$PIV = V_m = 10.61\text{ V}$$

The diode datasheets specify 1N4001 diode can meet these requirements. Chose 4 numbers of 1N4001 diodes.

1.3 VOLTAGE REGULATOR USING ZENER DIODE WITH FIXED LOAD CURRENT

Design Specifications

(*i*) DC output voltage = 5 V
(*ii*) Load current = 150 mA
(*iii*) Input voltage range = 10 to 15 V.

Theory

The circuit diagram of Zener voltage regulator is shown in Fig. 1.3. The voltage drop across the Zener terminals remains constant as the value of its resistance changes with the current. Thus, the Zener voltage V_Z is almost independent of the reverse diode current. The regulator circuit maintains almost constant output voltage even though the DC supply voltage and the load current may vary over a wide range. As the Zener diode is connected in shunt or parallel with the load R_L, the voltage regulator is known as a shunt regulator. The value of current-limiting resistance R_s should be such that the diode can operate in the breakdown region over the entire range of input voltage V_S and variations of the load current I_L.

If the supply voltage V_S varies, then the Zener current I_Z will vary because of the presence of r_Z, thereby causing a variation of the output voltage. This variation of the output voltage is defined by a factor called the line regulation, related to R_s and r_Z as

$$\text{Line regulation} = \frac{\Delta V_O}{\Delta V_S} = \frac{r_Z}{r_Z + R_s}$$

If the load current I_L increases, then the Zener current I_Z will decrease because of the presence of r_Z, thereby causing a decrease of the output voltage. This variation of the output voltage is defined by a factor called the load regulation, which is related to R_S and r_Z as

$$\text{Load regulation} = \frac{\Delta V_O}{\Delta I_L} = -\,r_Z \parallel R_s$$

Any change in the Zener voltage V_{ZO} will increase the output voltage. The variation of the output voltage is defined by a factor called the Zener regulation, given by

$$\text{Zener regulation} = \frac{\Delta V_O}{\Delta V_{ZO}} = \frac{R_s}{r_Z + R_s}$$

Thus, applying the superposition theorem, the effective output voltage V_O of the regulator is

$$V_O = \left(\frac{\Delta V_O}{\Delta V_{ZO}}\right)\Delta V_{ZO} + \left(\frac{\Delta V_O}{\Delta V_S}\right)\Delta V_S + \left(\frac{\Delta V_O}{\Delta I_L}\right)\Delta I_L$$

$$= \frac{R_s}{r_Z + R_s}\Delta V_{ZO} + \frac{r_Z}{r_Z + R_s}\Delta V_S - (r_Z \| R_s)\Delta I_L$$

Circuit Diagram

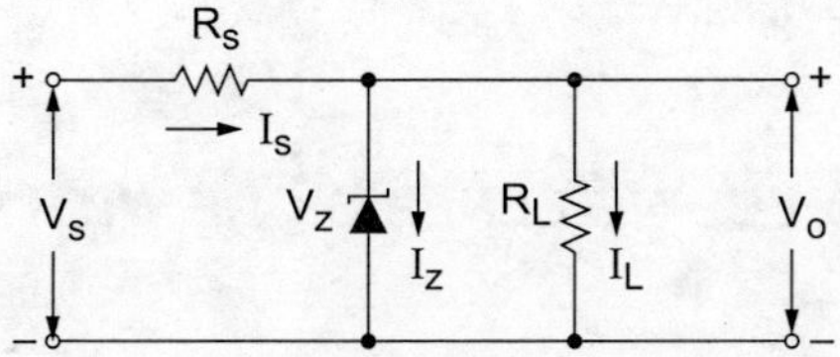

Fig. 1.3: Zener diode voltage regulator with fixed load

Design Equations

The minimum and maximum input voltages are V_{Smin} and V_{Smax}.

The Zener voltage, $V_z = V_{zo} + I_z \cdot r_z$

where V_{zo} is the value of V_z for $I_z = 0$ and r_z is Zener impedance at I_z.

Maximum Zener current, $I_{zmax} = \dfrac{P_{zmax}}{V_z}$

The series resistor $R_S \le \dfrac{V_{Smin} - V_{zo} - I_{zmin} r_z}{I_{zmin} + I_{Lmax}}$

where I_{zmin} and I_{Lmax} are the minimum Zener and maximum load currents, respectively.

For the value of R_S, the maximum Zener current flowing is

$$I_{zmax} = \frac{V_{Smax} - V_{zo} - I_{Lmin} R_S}{R_S + r_z}$$

The power dissipation in R_S,

$$P_{RS} = (V_{Smax} - V_Z)(I_{zmax} + I_{Lmin})$$

The worst-case power dissipation occurs in R_S when output is short-circuited.

Thus, $P_{RSmax} = \dfrac{V_{Smax}^2}{R_S}$

Load resistor, $R_L = \dfrac{V_O}{I_L}$

Design Steps

Chose a Zener diode 1N4733A having $V_z = 5.1$ V and $r_z = 7\ \Omega$ at $I_z = 49$ mA, $I_{zmin} = 1$ mA, and $P_{zmax} = 1$ W.

$$V_Z = V_O = 5.1 \text{ V}$$

$$V_{zo} = 4.757 \text{ V}$$

$$I_{Zmax} = 196 \text{ mA}$$

$$V_{Smin} = 10 \text{ V} \quad \text{and} \quad V_{Smax} = 15 \text{ V}$$

Series resistor $R_S \leq 34.67\ \Omega$

Choose $R_S = 33\ \Omega$

For this value of R_S, the maximum Zener current flowing is

$$I_{zmax} = 132.325 \text{ mA}$$

This is lower than the value found as per specifications.

The power dissipation in R_S, $P_{RS} = 2.795$ W

The worst-case power dissipation occurs in R_S when output is short-circuited. Thus,

$$P_{RSmax} = 6.82 \text{ W}$$

$$R_L = 34\ \Omega$$

1.4 VOLTAGE REGULATOR USING ZENER DIODE WITH VARIABLE LOAD CURRENT

Design Specifications

(*i*) DC output voltage = 7.5 V (*ii*) Load current = 0 to 100 mA

(*iii*) Input voltage range = 15 to 18 V.

The parameters of the Zener diode 1N4737A are $V_z = 7.5$ V and $r_z = 4\ \Omega$ at $I_z = 34$ mA, $I_{zmin} = 0.5$ mA, and $P_{zmax} = 1$ W.

Theory

The circuit diagram of Zener voltage regulator is shown in Fig. 1.4. The voltage drop across the Zener terminals remains constant as the value of its resistance changes with the current. Thus, the Zener voltage V_Z is almost independent of the reverse diode current. The regulator circuit maintains almost constant output voltage even though the DC supply voltage and the load current may vary over a wide range. As the Zener diode is connected in shunt or parallel with the load R_L, the voltage regulator is known as a shunt regulator. The value of current-limiting resistance R_s should be such that the diode can operate in the breakdown region over the entire range of input voltage V_S and variations of the load current I_L.

If the supply voltage V_S varies, then the Zener current I_Z will vary because of the presence of r_Z, thereby causing a variation of the output voltage. This variation of the output voltage is defined by a factor called the line regulation, related to R_s and r_Z as

$$\text{Line regulation} = \frac{\Delta V_O}{\Delta V_S} = \frac{r_Z}{r_Z + R_s}$$

If the load current I_L increases, then the Zener current I_Z will decrease because of the presence of r_Z, thereby causing a decrease of the output voltage. This variation of the output voltage is defined by a factor called the load regulation, which is related to R_S and r_Z as

$$\text{Load regulation} = \frac{\Delta V_O}{\Delta I_L} = -\, r_Z \,||\, R_s$$

Any change in the Zener voltage V_{ZO} will increase the output voltage. The variation of the output voltage is defined by a factor called the Zener regulation, given by

$$\text{Zener regulation} = \frac{\Delta V_O}{\Delta V_{ZO}} = \frac{R_s}{r_Z + R_s}$$

Thus, applying the superposition theorem, the effective output voltage V_O of the regulator is

$$V_O = \left(\frac{\Delta V_O}{\Delta V_{ZO}}\right)\Delta V_{ZO} + \left(\frac{\Delta V_O}{\Delta V_S}\right)\Delta V_S + \left(\frac{\Delta V_O}{\Delta I_L}\right)\Delta I_L$$

$$= \frac{R_s}{r_Z + R_s}\Delta V_{ZO} + \frac{r_Z}{r_Z + R_s}\Delta V_S - (r_Z \,||\, R_s)\Delta I_L$$

Circuit Diagram

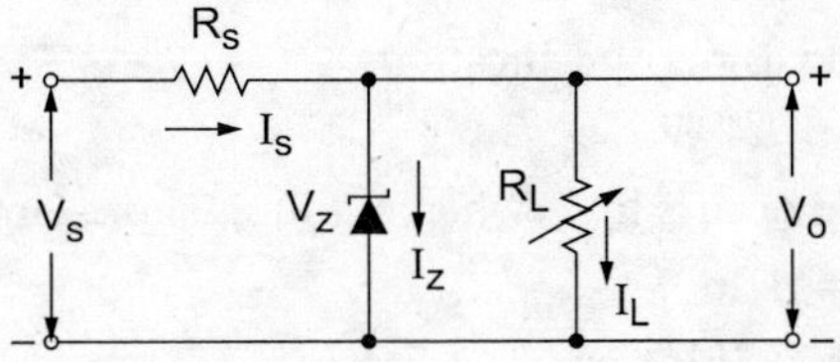

Fig. 1.4: Zener diode voltage regulator with variable load

Design Equations

The minimum and maximum input voltages are $V_{S\min}$ and $V_{S\max}$.

The Zener voltage, $V_z = V_{zo} + I_z r_z$

where V_{zo} is the value of V_z for $I_z = 0$ and r_z is Zener impedance at I_z.

Maximum Zener current, $I_{z\max} = \dfrac{P_{z\max}}{V_z}$

The series resistor $R_S \le \dfrac{V_{s\min} - V_{zo} - I_{z\min} r_z}{I_{z\min} + I_{L\max}}$

where $I_{z\min}$ and $I_{L\max}$ are the minimum Zener and maximum load currents, respectively.

For the value of R_S, the maximum Zener current flowing is,

$$I_{z\max} = \frac{V_{S\max} - V_{zo} - I_{L\min} R_S}{R_S + r_z}$$

The power dissipation in R_S

$$P_{RS} = (V_{S\max} - V_Z) \cdot (I_{z\max} + I_{L\min})$$

The worst-case power dissipation occurs in R_S when output is short-circuited.

Thus, $P_{RS\max} = \dfrac{V_{S\max}^2}{R_S}$

The minimum value of load resistor, $R_{L\min} = \dfrac{V_O}{I_{L\max}}$

The maximum value of load resistor, $R_{L\max} = \dfrac{V_O}{I_{L\min}}$

where $I_{L\min}$ is the minimum value of load current.

Design Steps

$$V_Z = V_O = 7.5 \text{ V}$$

$$V_{zo} = 7.364 \text{ V}$$
$$I_{Zmax} = 133.33 \text{ mA}$$
$$V_{Smin} = 15 \text{ V} \quad \text{and} \quad V_{Smax} = 18 \text{ V}$$

Series resistor $R_S \leq 76\ \Omega$

Choose $R_S = 76\ \Omega$

For this value of R_S, the maximum Zener current flowing is

$$I_{zmax} = 132.95 \text{ mA}$$

This is lower than the value found as per specifications.

The power dissipation in R_S, $P_{RS} = 1.4$ W

The worst-case power dissipation occurs in R_S when output is short-circuited.

Thus, $P_{RSmax} = 4.26$ W

$$R_{Lmin} = 75\ \Omega$$
$$R_{Lmax} = \infty$$

1.5 PARALLEL CLIPPER

Design Specifications

(*i*) Input signal $V_i = 5 \sin 2x\, 10^3\, \pi t$

(*ii*) Clipping level = above + 2.5 V

(*iii*) The diode forward voltage drop $V_D = 0.7$ V

(*iv*) Maximum current through the diode = 20 mA.

Theory

A clipper in which the diode is connected across the output terminals is known as a parallel or shunt clipper because the diode will be in parallel or shunt with the load. In a shunt connection, elements are connected in parallel such that each element carries a different current. A shunt clipper circuit and the corresponding output waveform for sinusoidal input are shown in Fig. 1.5. The resistance *R* limits the diode current when the diode conducts. In determining the output waveform of a clipper, it is important to keep in mind that a diode will conduct only if the anode voltage is higher than the cathode voltage.

When diode *D* is OFF, the instantaneous output voltage V_O equals the instantaneous input voltage V_i. Diode *D* will conduct for the portion of the positive half-cycle during which the instantaneous input voltage $V_i > (V_R + V_D)$.

Circuit Diagram and Waveforms

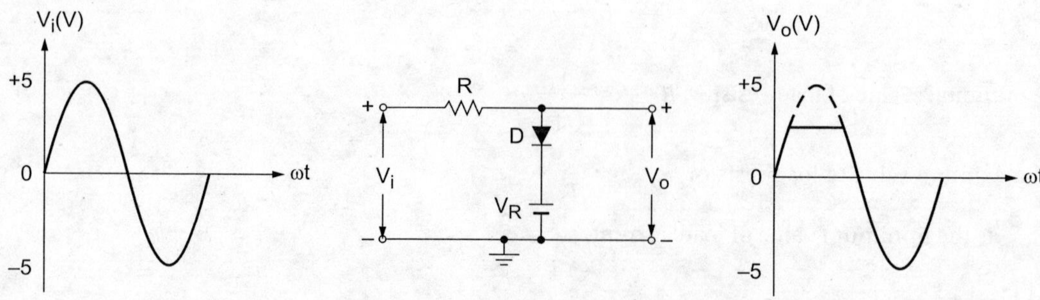

Fig. 1.5: Parallel clipper with input and output voltage waveforms

Design Steps

A shunt diode clipper with a battery V_R as shown in figure can be used. In order to clip the input signal above + 2.5 V, $V_D + V_R = 2.5$ V or $V_R = 1.8$ V.

The current through diode, $I = \dfrac{V_i - 2.5}{R}$

$$R = \frac{2.5\text{ V}}{20\text{ mA}} = 125\ \Omega$$

Chose 1N4001 diode.

1.6 SERIES CLIPPER

Design Specifications

(*i*) Input V_i = sinusoidal signal of peak amplitude 5 V and frequency 1 kHz

(*ii*) The diode forward voltage drop $V_D = 0.7$ V

(*iii*) Maximum current through the diode = 20 mA.

Theory

A clipper in which the diode forms a series circuit with the output terminals is known as a series clipper. A series clipper circuit and the corresponding output waveform for sinusoidal input are shown in Fig. 1.6. The current limiting resistance *R* can be used as a load. In determining the output waveform of a clipper, it is important to keep in mind that a diode will conduct only if the anode voltage is higher than the cathode voltage.

The potential difference across the battery must be V_R. The negative terminal of the battery cannot be at zero or ground potential. Therefore, these circuits require an isolated DC voltage or battery of V_R. Note that the zero level of the output voltage V_O is different from that of the input voltage V_i and is shifted by an amount equal to V_R.

The diode is forward biased for input voltage $V_i < (V_R - V_D)$ and the output voltage $V_O = V_i - (V_R - V_D)$. For $V_i > (V_R - V_D)$, the diode is OFF and the output is zero.

Circuit Diagram and Waveforms

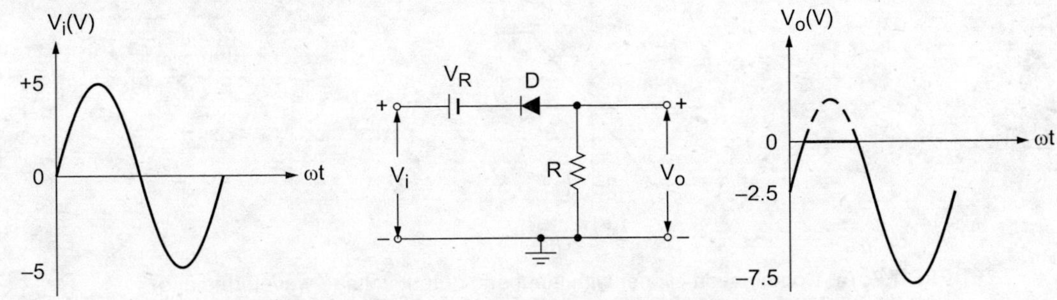

Fig. 1.6: Series clipper with input and output voltage waveforms

Design Steps

A series diode clipper with a battery V_R as shown in Fig. (*b*) can be used. In order to get the required output, $V_R - V_D = 2.5$ V or $V_R = 3.2$ V. The current through diode, $I = \dfrac{7.5V}{R}$.

$$R = 375\ \Omega$$

Chose 1N4001 diode.

1.7 COMBINATION CLIPPER

Design Specifications

(*i*) Input V_i = sinusoidal signal of peak amplitude 5 V and frequency 1 kHz

(*ii*) The diode forward voltage drop $V_D = 0.7$ V

(*iii*) Maximum current through the diode = 20 mA.

Theory

The circuit shown in Fig. 1.7(*a*) is a double clipper using shunt diodes and reference voltage sources. It is the combination of positive and negative clippers. During positive half-cycle of input voltage V_i and for $V_i > (V_{D1} + V_{R1})$, D_1 is forward biased and D_2 is OFF. The output voltage $V_O = (V_{D1} + V_{R1})$. For $V_i < -(V_{D2} + V_{R2})$ the diode D_1 is OFF, D_2 is forward biased and the output voltage $V_O = -(V_{D2} + V_{R2})$. For $-(V_{D2} + V_{R2}) < V_i < (V_{D1} + V_{R1})$, both the diodes are OFF and $V_O = V_i$, as shown in Fig. 1.7(*b*).

Circuit Diagram and Waveforms

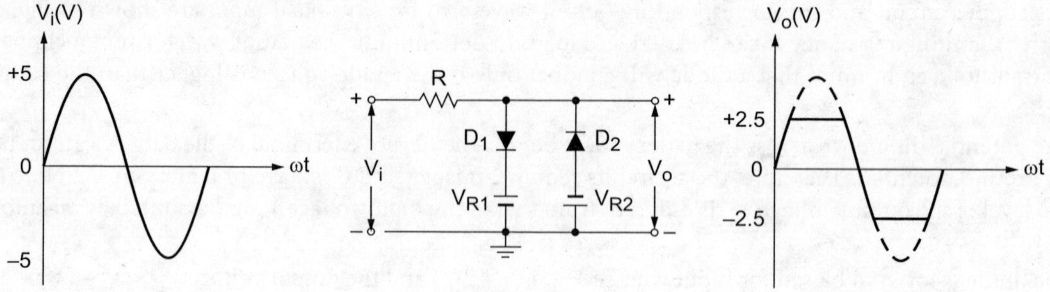

(*a*)

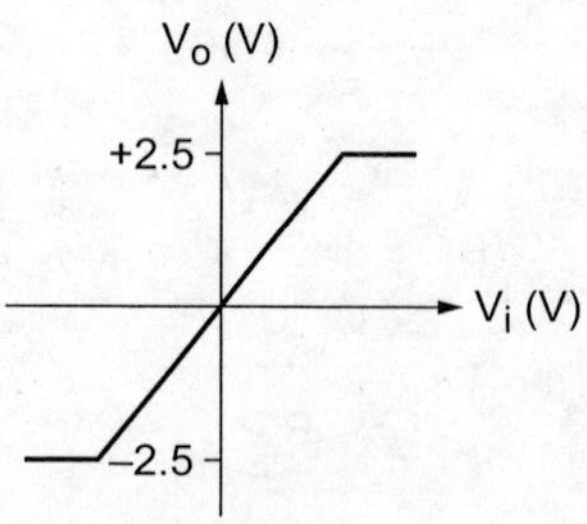

(*b*)

Fig. 1.7: (a) Combination clipper with input and output voltage waveforms and (b) voltage transfer characteristics

Design Steps

To obtain the required transfer characteristics, a double clipper using shunt diodes and batteries as shown in Fig. 1.7 (*b*) can be used. In order to clip the positive half-cycle of the input signal above + 2.5 V, $V_{D1} + V_{R1}$ = 2.5 V or $V_{R1} = 2.5 - 0.7 = 1.8$ V.

For the negative half-cycle of the input signal to be clipped below – 2.5 V, $V_{D2} + V_{R2} = 2.5$ V or $V_{R2} = 2.5 - 0.7 = 1.8$ V.

During positive half-cycle of the input, diode D_1 conducts and during negative half-cycle the diode D_2 conducts. To limit the current through each of the diodes to 20 mA, the series resistor R must be $\geq (5 - 2.5)\text{V}/20\text{mA} = 125\ \Omega$.

Chose 1N4001 diodes for D_1 and D_2.

1.8 FUNCTION GENERATOR CIRCUIT FOR THE VOLTAGE TRANSFER CHARACTERISTICS

The required piece-wise linear voltage transfer characteristics of a diode network are shown in Fig. 1.8.

Theory

The piece-wise linear voltage transfer characteristics shown in Fig. 1.8 can be obtained using resistor, diode, and battery network. The diode conducts when it is forward biased, i.e., when its anode is at a higher potential than the cathode. Different slopes can be obtained using resistor network. The batteries provide reference voltages that will control the ON-OFF states of diodes.

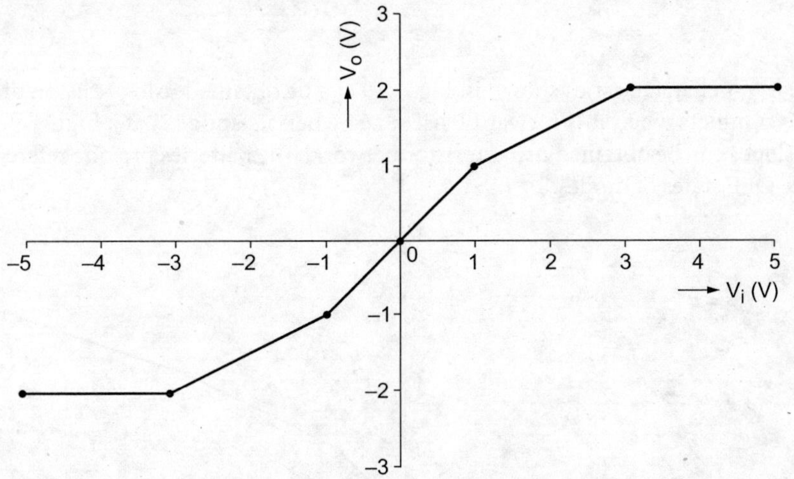

Fig. 1.8: Piece-wise linear voltage transfer characteristics of a diode network

Design Steps

Let us assume that the diodes are ideal.

The input voltage V_i in the circuit diagram is the variable DC voltage and is varied from –5 V to +5 V.

Positive V_i: Initially, for $V_i = 1$ V, $V_o = 1$ V, *i.e.*, slope = 1. A series resistor R_1 of value, say 1 kΩ can be taken. When V_i changes from 1 to 3 V, V_o changes from 1 to 2 V, giving a slope of 1/2. For this part of the characteristic, a resistor R_3, diode D_3, and a battery of 1 V, all in series are to be used as shown in the circuit.

$$V_o = \frac{V_i R_3}{R_1 + R_3} \quad \text{or} \quad \frac{R_3}{R_1 + R_3} = \frac{1}{2}$$

$$R_3 = 1\ \text{k}\Omega$$

For the third segment and for $V_i > 3$ V, V_o is constant at 2 V and hence the slope = 0. A diode D_4 and a battery of 2 V connected in series are to be used as shown in the circuit given in Fig. 1.9.

Negative V_i: Since the characteristics are similar, the same circuit with diodes D_1 and D_2, Resistor R_2, and batteries 2 V and 1 V reversed are used.

Circuit Diagram

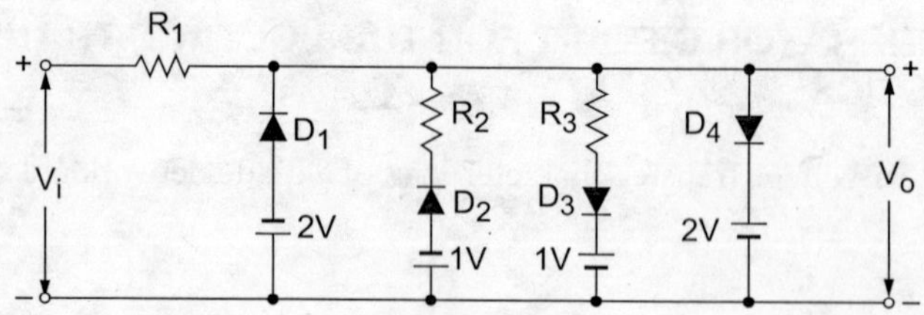

Fig. 1.9: Circuit diagram for the voltage transfer characteristics of Fig. 1.8.

$R_1 = R_2 = R_3 = 1$ kΩ and chose 1N4001 diodes for $D_1 - D_4$.

1.9 FUNCTION GENERATOR CIRCUIT FOR THE V-I CHARACTERISTICS

Theory

The piece-wise linear *V-I* characteristics shown in Fig. 1.10 can be obtained using resistor, diode, and battery network. The diode conducts when it is forward biased, i.e., when its anode is at a higher potential than the cathode. Different slopes can be obtained using resistor network. The batteries provide reference voltages that will control the ON-OFF states of diodes.

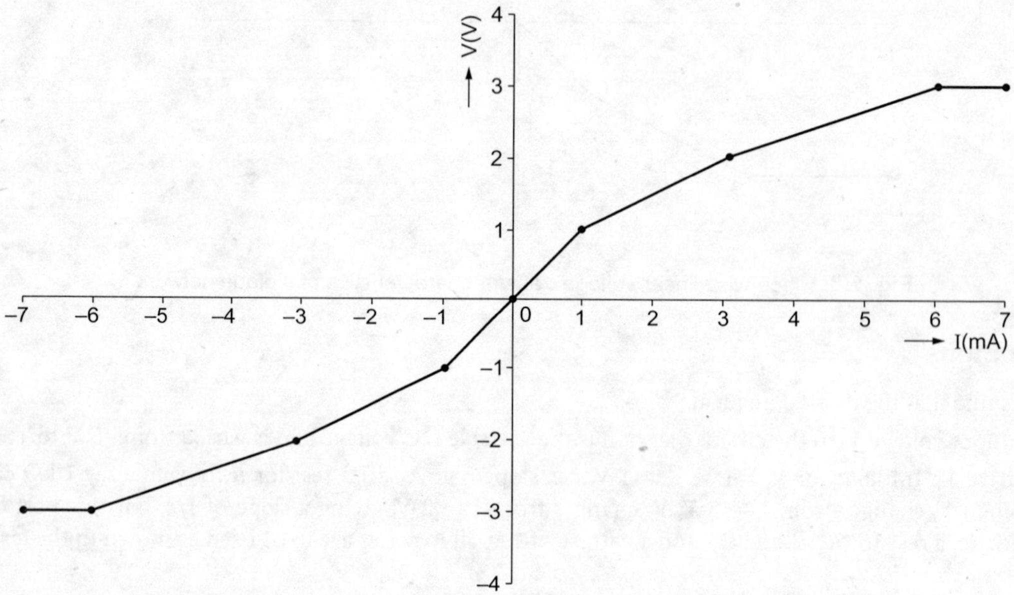

Fig. 1.10: Piece-wise linear V-I characteristics of a diode network

Design Steps

Let us assume that the diodes are ideal.

A resistor R of, say, 1 kΩ in series with a variable DC supply V_i of 10 V, varied from −10 V to +10 V, is used to get different currents. The voltage V is the output voltage V_o as shown in the circuit diagram.

Positive V: Initially, for $V = 1$ V, $I = 1$ mA, *i.e.*, slope of 1 kΩ, use a shunt resistor $R_1 = 1$ kΩ. Now, as soon as the voltage V reaches a value of 1 volt, the slope changes to $\frac{(2-1)\text{V}}{(3-1)\text{mA}}$, *i.e.*, a resistance of 0.5 kΩ. Now use another shunt resistor R_2, such that $R_1 \,||\, R_2 = 0.5$ kΩ. This gives R_2 as 1 kΩ. Since the slope change occurs above 1V, use a diode and battery of 1V in series with R_2 as shown in the circuit diagram. Similarly, for V between 2 and 3V, the slope is $\frac{(3-2)\text{V}}{(6-3)\text{mA}}$, *i.e.*, a resistance of $\frac{1}{3}$ kΩ. Use another shunt resistor R_3, such that $R_1 \,||\, R_2 \,||\, R_3 = \frac{1}{3}$ kΩ. This gives R_3 as 1kΩ. A diode and a battery of 2V are to be connected in series with R_3. For V above 3V, slope is zero. So, only diode and a battery of 3V connected in series are to be used as shown in Fig. 1.11.

Negative V: Since the characteristics are similar, the same circuit with diodes and batteries reversed are used.

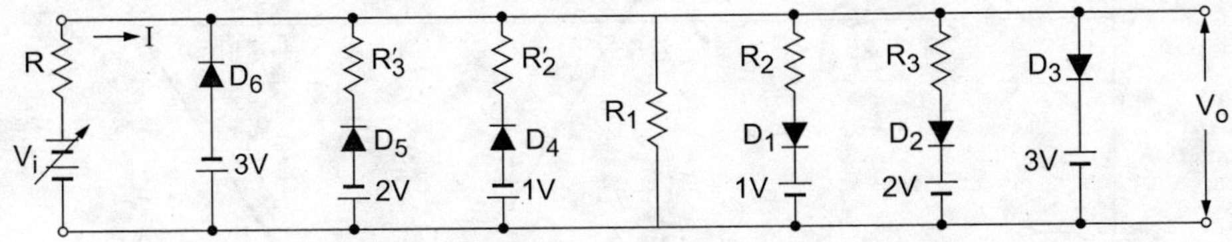

Fig. 1.11: Circuit diagram for the V-I characteristics of Fig. 1.10.

$R = R_1 = R_2 = R_3 = R_2' = R_3' = 1$ kΩ and chose 1N4001 diodes for $D_1 - D_6$.

1.10 WAVEFORM GENERATOR THAT CONVERTS A TRIANGULAR WAVE INTO A SINUSOIDAL WAVE

Theory

The given waveform shown in Fig. 1.12 indicates a non-linear function. By looking at the input and required output waveforms, it can be noted that the triangle will remain a triangle up to a certain point; then, it gets attenuated steeply by bringing in large attenuation factors. This can be accomplished by the diode transfer function generator.

Design Steps

Let us assume that the diodes are ideal.

The given task can be accomplished by means of diode function generator whose transfer characteristics are shown in Fig. 1.13. The input triangular wave V_i is indicated as a solid line and the required output V_o is indicated as a dotted line. For $V_i = 0$ to 2.5 V, the output follows the input. When input voltage increases from 2.5 to 5 V, the output increases to a maximum value of 4 V. During the falling edge, the output decreases from 4 to 2.5 V and thereafter follows input till the input reaches – 2.5 V. For input changes from – 2.5 to – 5 V, the output falls to a minimum of – 4 V. Again, when input rises from – 5 V to 0 V, the output increases from – 4 V to – 2.5 V and then follows the input.

Positive V_i: First segment has a slope of 1, *i.e.*, $\frac{\Delta V_o}{\Delta V_i} = 1$. Insert a series resistor R_1 of say 1 kΩ. The second segment has a slope of $\frac{1.5\text{ V}}{2.5\text{ V}} = 0.6$. A resistor R_2, diode D_1 and a battery of 2.5 V, all in series to be used as shown in the circuit diagram.

$$0.6 = \frac{R_2}{R_1 + R_2}$$

$$R_2 = 1.5\ \text{k}\Omega$$

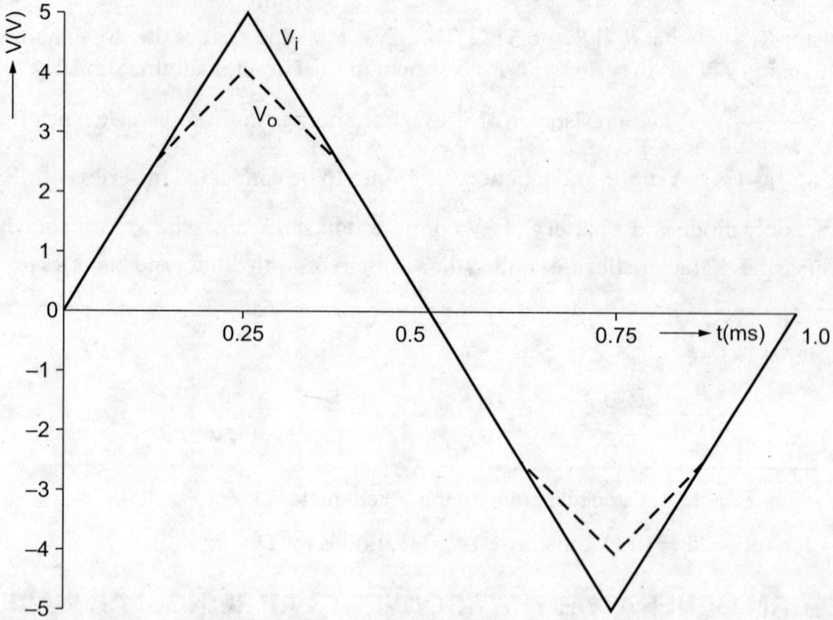

Fig. 1.12: Triangular input waveform (solid line) and the required sinusoidal waveform (dotted line)

Negative V_i: For the negative portion, since the slope is same, a resistor R_3 equal to R_2 is used in series with changed diode and battery combination as shown.

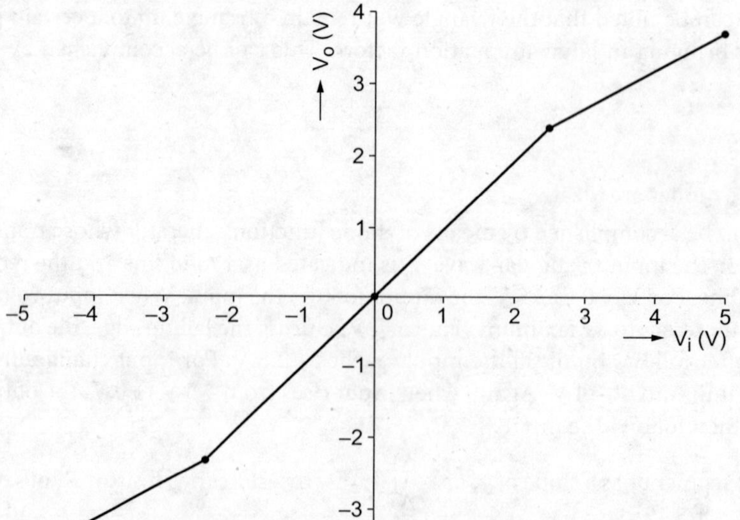

Fig. 1.13: Diode function generator voltage transfer characteristics corresponding to waveforms shown in Fig. 1.12.

Circuit Diagram

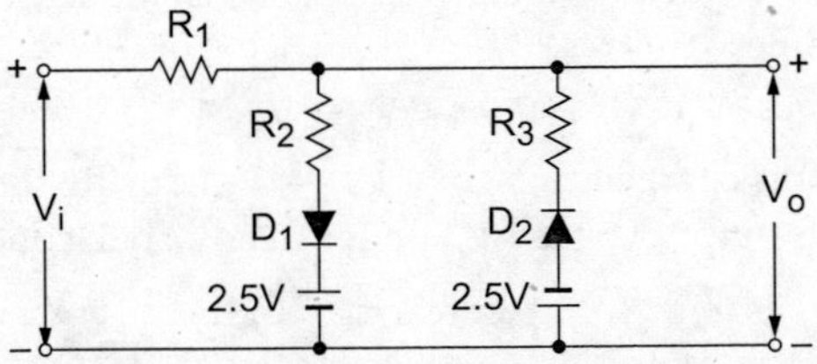

Fig. 1.14: Circuit diagram for the voltage transfer characteristics of Fig. 1.13.

$R_1 = 1\ k\Omega$, $R_2 = 1.5\ k\Omega$, $R_3 = 1.5\ k\Omega$, and chose 1N4001 diodes for D_1 and D_2.

1.11 CLAMPER

Design Specifications

(*i*) Input signal - symmetrical square wave of peak-to-peak amplitude 10 V and frequency 1 kHz

(*ii*) Required output clamping levels: (*a*) positive peak at + 1 V, (*b*) positive peak at – 2 V, (*c*) negative peak at 0 V, (*d*) negative peak at – 3 V

(*iii*) Assume ideal diodes.

Theory

A clamping circuit, also known as a DC restorer, changes the DC voltage level of a waveform, but does not affect its shape. A clamping network must have a capacitor, a diode and a resistive element. The magnitude *R* and *C* must be chosen such that the time constant RC is large enough to ensure that the voltage across the capacitor does not discharge significantly during the interval the diode is non-conducting. By connecting suitable DC voltage in series with the diode, the level of swing can be varied. The clamping circuit which shifts the signal vertically upwards is called a positive clamper. The negative clamper does the reverse, that is, it pushes the signal downwards so that the positive peak falls on the zero level.

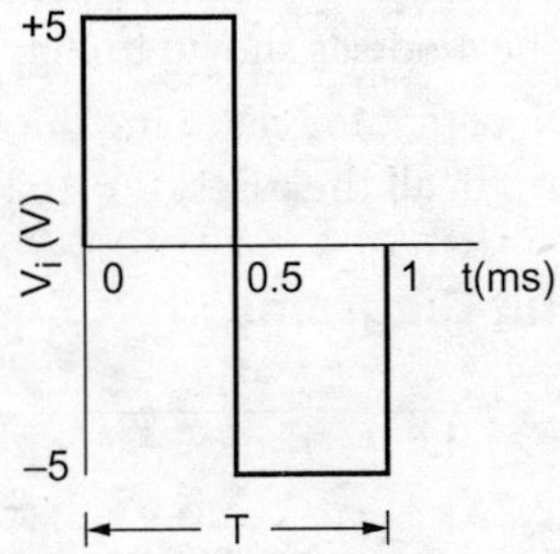

Fig. 1.15: Input voltage waveform

The input signal, which is a symmetrical square wave is shown in Fig. 1.15. The output voltage V_O can be shifted to a predefined value by introducing a battery voltage V_R. Consider the clamping circuit shown in Fig. 1.16(*a*). If V_m is the maximum or peak value of input voltage, then the capacitor *C* will be charged to $V_C = V_m - V_R$ during the first positive half-cycle of the input voltage. The instantaneous output voltage V_O under steady-state conditions becomes,

$$V_O = V_i - V_C = V_i - (V_m - V_R) = V_i - V_m + V_R$$

The capacitor *C* in Fig. 1.16(*b*) will be charged to $V_C = V_m + V_R$ during the first positive half-cycle of the input voltage and the instantaneous output voltage V_O under steady-state conditions is

$$V_O = V_i - V_C = V_i - (V_m + V_R) = V_i - V_m - V_R$$

In the circuit shown in Fig. 1.16(*c*), the diode *D* will be reverse biased during the first positive half-cycle of the input voltage and the output voltage will be equal to the input voltage, $V_O = V_i$. The diode *D* will conduct during the first negative half-cycle of the input voltage, the capacitor *C* will be charged almost instantaneously to the negative peak input voltage – V_m, and the output voltage V_O will become zero. This process is completed during the first cycle, and the circuit reaches a steady-state condition with an input voltage of – V_m across the capacitor C. After the first cycle, the capacitor voltage remains constant at $V_C = -V_m$. The output voltage V_O under steady-state conditions becomes

$$V_O = V_i - V_C = V_i - (-)V_m = V_i + V_m$$

The capacitor C in Fig. 1.16(*d*) is charged to $V_C = -(V_m - V_R)$ during the first negative half-cycle of the input voltage and the instantaneous output voltage V_O under steady-state conditions is

$$V_O = V_i - V_C = V_i - [-(V_m - V_R)] = V_i + V_m - V_R$$

Design Steps:

For proper operation of the circuit, RC must be much larger than T.

Chose, $RC = 10\,T = 10$ ms

If R is taken as 10 kΩ, then $C = 1$ μF

Chose 1N4001 diodes.

In order to clamp the positive peak of the signal at + 1V, a battery voltage V_R of 1V is to be connected in series with the diode as shown in Fig. 1.16(*a*).

To clamp the positive peak of the signal at – 2V, a battery voltage V_R of 2V is to be connected in series with the diode as shown in Fig. 1.16(*b*).

For clamping the negative peak of the signal at 0V, no battery is required to be connected in series with the diode as shown in Fig. 1.16(*c*).

To clamp the negative peak of the signal at – 3V, a battery voltage V_R of 3V is to be connected in series with the diode as shown in Fig. 1.16(*d*).

Note that for clamping the negative peaks, the diode terminals are reversed as shown in Figs. 1.16(*c*) and 1.16(*d*). In all the cases, $R = 10$ kΩ and $C = 1$ μF.

Circuit Diagrams

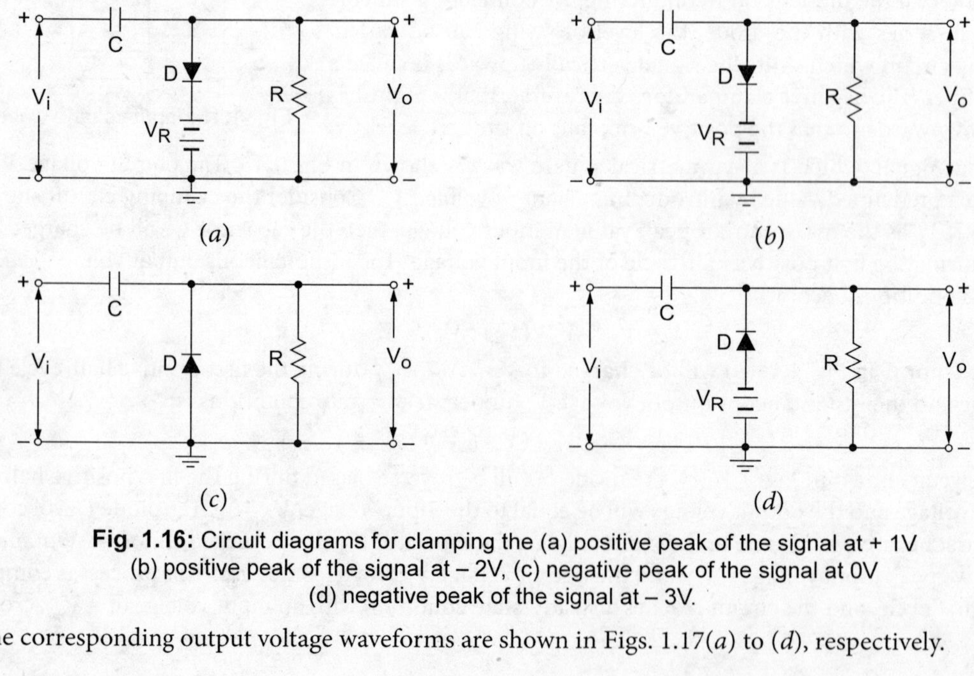

Fig. 1.16: Circuit diagrams for clamping the (a) positive peak of the signal at + 1V (b) positive peak of the signal at – 2V, (c) negative peak of the signal at 0V (d) negative peak of the signal at – 3V.

The corresponding output voltage waveforms are shown in Figs. 1.17(*a*) to (*d*), respectively.

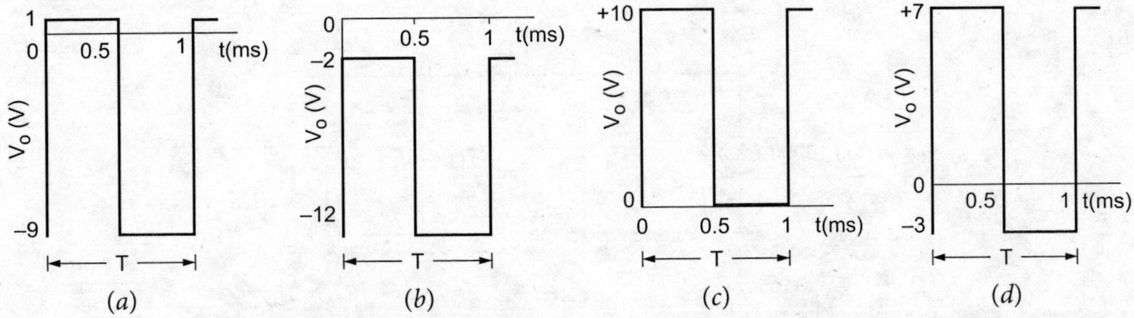

Fig. 1.17: Output voltage waveforms of circuit diagrams shown in Fig. 1.16.

1.12 VOLTAGE DOUBLER

Design Specifications

(*i*) Input-Sinusoidal signal of $V_M = 5$ V and frequency $f = 50$ Hz

(*ii*) Output voltage $\approx 2\ V_M = 10$ V (*iii*) Output current = 10 mA

(*iv*) Average peak-to-peak ripple voltage for both sections shall not exceed 5% of output voltage.

Theory:

The voltage doubler circuit shown in Fig. 1.18 uses a clamping circuit and a peak detector. The input is $V_i = V_M \sin \omega t$. The circuit operation can be divided into four intervals: interval 1, interval 2, interval 3, and interval 4.

During interval 1, *i.e.*, $0 \le \omega t \le \pi/2$, the diode D_1 will conduct, but diode D_2 will be reverse biased. The output voltage is $V_O = 0$. The capacitor C_1 will be charged during the first quarter-cycle to V_M at $\omega t = \pi/2$ with the polarities shown.

During interval 2, *i.e.*, $\pi/2 \le \omega t \le \pi$, the diode D_1 will be OFF and D_2 is ON. If the capacitor C_1 will not have time to discharge, the voltage on capacitor C_1 will remain approximately at V_M.

Interval 3 is the interval during which $\pi \le \omega t \le 3\pi/2$. The polarity of the input voltage is negative and hence diode D_1 will be OFF and diode D_2 will conduct. The output voltage V_O, which will be the same as the voltage across capacitor C_2, will become,

$$V_O = V_{C1} - V_i = V_M - V_M \sin \omega t$$

At $\omega t = 3\pi/2$, the output voltage is $2V_M$ and the capacitor C_2 will be charged to $2V_M$.

During interval 4, *i.e.*, during $3\pi/2 \le \omega t \le 2\pi$, the diodes D_1 and D_2 will be OFF. The voltage on capacitor C_1 will be $V_{C1} = V_M$, and that on capacitor C_2 will be $V_{C2} = 2V_M$. However, the capacitor C_1 acts as the voltage source of V_M and contributes to charging C_2. In fact, C_1 and C_2 form a series circuit and share $2V_M$, so the voltage on capacitor C_2 will be less than $2V_M$. It will take about two cycles before the steady-state condition is reached.

Circuit Diagram and Waveforms

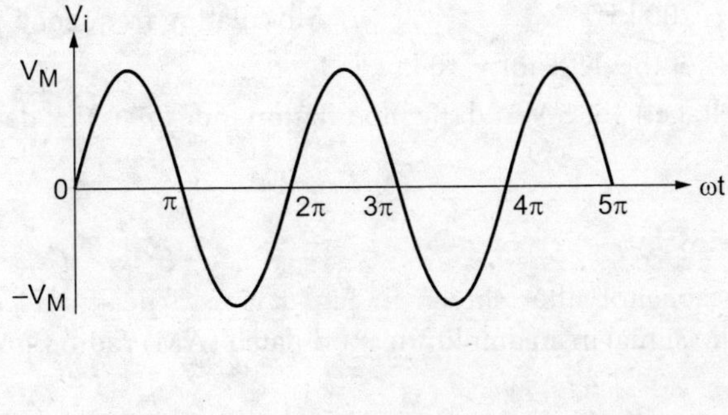

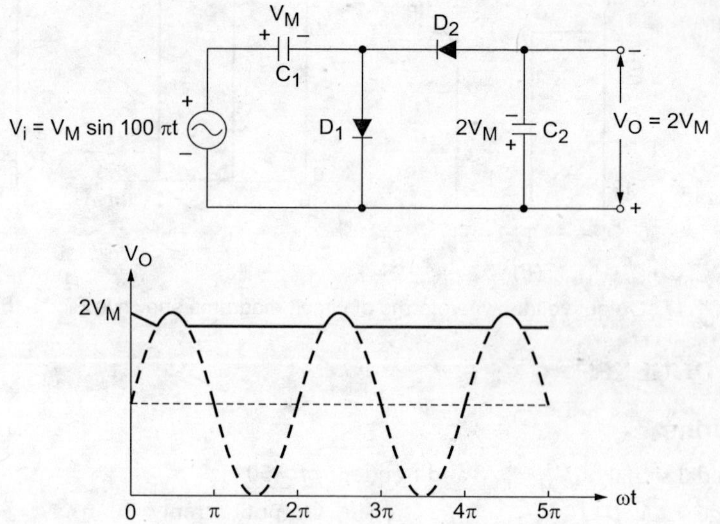

Fig. 1.18: Voltage doubler circuit and input and output voltage waveforms

Design Equations

Average ripple voltage is given by

$$V_{rav} = \left[\frac{I_o}{4fC}\right] n(n+1)$$

where I_o is the output current, f is input signal frequency, C is capacitor, and n is the number of half-wave rectifier (HWR) stages in the voltage doubler circuit.

$$C = \frac{I_o n(n+1)}{4fV_{rav}}$$

Design Steps:

Given, $V_{rav} = 0.05V_O = 0.5$ V

As shown in the circuit, the voltage doubler consists of 2 HWR stages. Hence $n = 2$.

$$C = 600\ \mu\text{F}$$

Chose $C_1 = C_2 = 620\ \mu$F, 15 V and 1N4001 diodes.

1.13 AM DEMODULATOR

Design Specifications

(*i*) Carrier frequency, $f_c = 200$ kHz (*ii*) Modulating frequency, $f_m = 1$ kHz

(*iii*) The load resistance R_L of the detector = 10 kΩ.

If the peak modulating voltage V_m is 5 V and the modulation index m is 0.5, determine the amplitude and frequencies of sidebands.

Theory

The half-wave rectifier with capacitor filter shown in Fig. 1.19 is a peak detector and can be used as a demodulator to detect the audio signal in an amplitude modulated (AM) radio signal. Amplitude modulation is a method of translating a low-frequency signal into a high frequency one. The AM waveform can be described by

$$V_S(t) = V_m[1 + M \sin(2\pi f_m t)] \sin(2\pi f_c t)$$

where f_c = carrier frequency, in Hz

f_m = modulating frequency, in Hz

M = modulation index, whose value varies between 0 and 1

V_m = peak modulating voltage

The term $V_m[1 + M \sin(2\pi f_m t)]$ represents the envelope of the modulated waveform.

During the first quarter-cycle, the input voltage will rise, the capacitor C will be charged almost instantaneously to the input voltage, and the capacitor or output voltage V_O will follow the input voltage V_S until the instantaneous V_S reaches V_m at time $t = \pi/2\omega$. When the input voltage V_S tries to decrease, diode D will be reverse biased, and the capacitor C will discharge through resistance R_L.

A low-pass filter can be used to smooth the demodulated signals. With a proper choice of time constant $\tau = RC$, the output will trace each peak of the modulating signal. If the time constant is too large, the output will not be able to change fast enough and the audio signal will be distorted. If the time constant is too small, there will be too much ripple superimposed on the modulating signal.

Circuit Diagram

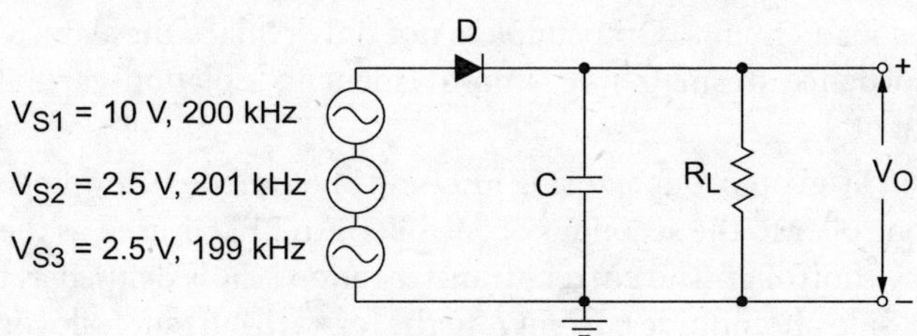

Fig. 1.19: AM demodulator

Design Equations

The AM waveform is denoted as

$$v_S(t) = V_c[1 + m \sin(2\pi f_m t)] \sin(2\pi f_c t)$$

where V_c is the peak amplitude of the carrier and $m = \dfrac{V_m}{V_c}$ or $V_c = \dfrac{V_m}{m}$

A peak detector will serve the purpose of detecting the AM wave.

$$C = \frac{1}{2\pi f_m R_L}$$

$$\text{Amplitude of the sideband} = \frac{mV_c}{2}$$

$$f_{USB} = f_c + f_m$$

$$f_{LSB} = f_c - f_m$$

Design Steps

$$V_c = 10 \text{ V}$$

$$C = 15.92 \text{ nF}$$
$$\text{Amplitude of the sideband} = 2.5 \text{ V}$$
$$f_{USB} = 201 \text{ kHz}$$
$$f_{LSB} = 199 \text{ kHz}$$

Chose 1N4148, a small-signal fast switching diode for the application.

1.14 OPTOCOUPLER USING LED AND PHOTODIODE

Design Specifications:

(*i*) Maximum current through the LED = 20 mA
(*ii*) The output voltage across the photodiode = 2.5 V.

Determine the current transfer ratio (CTR).

Theory

Optocouplers optically links, via transparent isolating material, a light emitter and a photodetector. Used as an interface between circuits with different ground potentials, optocouplers replace isolation transformers and electromagnetic relays. Traditionally, relays or transformers have been used for isolation interfaces between logic circuits and power line load circuits. Optocouplers not only replace these devices, but also have merits such as elimination of impedance mismatching, improvement in isolation capability between input and output, and ease of noise cut off.

Fig. 1.20 shows the circuit of an optocoupler using LED and photodiode. The optocoupler is used to transmit signal from primary side controller to the secondary controller. An LED is used as the primary controller and photodiode as the secondary controller. The current transfer ratio, CTR is defined as the ratio I_2/I_1, expressed as a percentage of the output side photodiode current I_2 to the forward current I_1 flowing in the input side LED.

$$\text{CTR} = \left(\frac{I_2}{I_1}\right) \times 100\%$$

Circuit Diagram

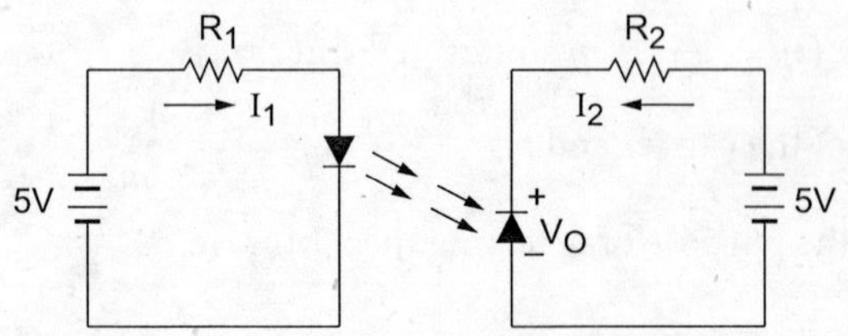

Fig. 1.20: Optocoupler using LED and photodiode

Design Equations

The resistor R_1 at the primary side controller is

$$R_1 = \frac{V_P - V_F}{I_F}$$

where V_P is the primary side controller supply voltage, V_F is LED forward voltage, and I_F is LED forward current.

The resistor R_2 at the secondary side controller is

$$R_2 = \frac{V_S - V_O}{I_R}$$

where V_S is the secondary side controller supply voltage, V_O is output voltage across the photodiode, and I_R is reverse light current of photodiode.

$$\text{Current transfer ratio, CTR} = \left(\frac{I_2}{I_1}\right) \times 100\%$$

where $I_1 = \dfrac{V_P - V_F}{R_1}$ is the current through the primary side controller and $I_2 = \dfrac{V_S - V_O}{R_2}$ is the current through the secondary side controller.

Design Steps

Chose infrared LED IR323 and BPV10NF photodiode.

From the datasheets, for 20 mA forward current, typical forward voltage of LED is 1.2 V.

$$R_1 = 190\ \Omega.$$

Typical reverse light current of photodiode for an irradiance of 1 mW/cm^2, λ = 870 nm, and V_R = 5 V is 55 μA.

$$R_2 = 45.45\ \text{k}\Omega$$

Current transfer ratio, CTR = 0.275%

❑❑❑

2

CHAPTER

MOSFET CIRCUITS

MOSFET CIRCUITS

The metal oxide semiconductor field effect transistor or MOSFET is a semiconductor device having four terminals, namely, source (S), gate (G), drain (D) and body (B). Generally, the source and body are tied together, forming a three-terminal device. The operation of the device depends on the flow of charge carriers, either electrons or holes, along the length of the channel, lying beneath the oxide layer, and in between source and drain terminals. The current through the channel can be controlled by the voltage applied to the gate terminal.

The MOSFETs are of two types, *N*-channel MOSFET and *P*-channel MOSFET. In the *N*-channel MOSFET, the channel is of N-type, source and drain regions are heavily doped N-type semiconductors, whereas the substrate or body is of P-type. The MOSFET can function in two modes, depletion and enhancement. The depletion mode MOSFET is normally ON device and the application of appropriate gate voltage causes depletion of charge carriers in the channel, causing the device go to OFF state. The enhancement mode MOSFET is normally OFF device and the application of appropriate gate voltage causes increase of charge carriers in the channel, causing the device switch to ON state. MOSFETs are widely used as switches and for amplification of electrical signals. They are employed in both analog and digital circuits.

In this chapter, the design of, switch circuit, voltage-divider bias circuit, common-source amplifier, source follower, cascode amplifier, differential amplifier, and simple current mirror, for given specifications, using N-channel enhancement MOSFET is given.

Note: Unless otherwise specified, N-channel enhancement MOSFET 2N7000 is used for the design of circuits. The device parameters are $k_n = 180\ \text{mA/V}^2$ and $V_t = 2.1$ V.

2.1 SWITCH CIRCUIT

Assume $\lambda = 0$ and $V_{DD} = 5$ V.

Theory

The maximum drain to source voltage rating determines the maximum voltage that can be switched. The gate threshold voltage determines the minimum voltage to be applied to the gate to make the MOSFET conduct. Enhancement MOSFETs when ON allow current in either direction with an essentially identical R_{DS}(ON). When OFF they block current in one direction. A MOSFET only requires gate current during the switching to charge the gate-source capacitance. This gate current can be high.

Figure 2.1 shows an arrangement to operate the *N*-channel enhancement MOSFET as a switch. Whenever the manual switch *S* is closed, 5 V is applied to the gate terminal through R_{G1}, the MOSFET conducts and the

LED glows. This is the ON state of the MOSFET switch. Whenever S is open, the MOSFET is OFF, no drain current flows and LED will not glow.

Circuit Diagram

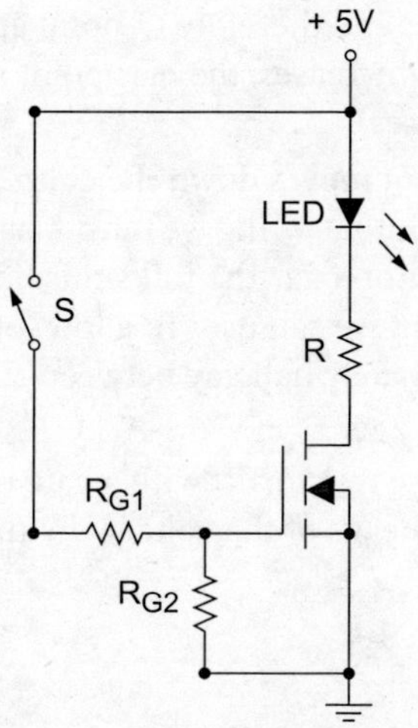

Fig. 2.1: MOSFET switch circuit

Design Steps

The input voltage shall be of sufficient amplitude to drive the transistor into ON state. A voltage of +5 V can be chosen. The ON and OFF states of the transistor are controlled by the switch S and are indicated by the light output or no light output of LED. Chose infrared LED IR323 having typical forward voltage of 1.2 V for 20 mA forward current.

Chose $R = 330\ \Omega$.

To prevent any ringing oscillations caused by the lead inductance and gate capacitance, a resistor R_{G1} is connected in series with the gate terminal of MOSFET.

Chose $R_{G1} = 100\ \Omega$.

In order to avoid floating of MOSFET gate terminal, a large value resistor R_{G2} is connected between gate and source terminals.

Chose $R_{G2} = 470\ \text{k}\Omega$.

2.2 VOLTAGE DIVIDER BIAS CIRCUIT

Design Specifications

(*i*) Supply voltage $V_{DD} = 20$ V (ii) Quiescent drain current $I_{DQ} = 30$ mA.

Assume $\lambda = 0$ for the 2N7000 MOSFET.

Theory

Figure 2.2 shows a voltage divider bias circuit for the N-channel enhancement MOSFET. Enhancement MOSFET is a normally OFF device, that is, it only conducts when a suitable positive gate-to-source voltage is applied. However, due to the construction and physics of an enhancement type MOSFET, there is a minimum gate-to-source voltage, called the threshold voltage V_{th} must be applied to the gate before it starts to conduct and allow drain current to flow. As the gate's forward bias increases, the drain current I_D will also increase.

For a MOSFET to operate as a linear amplifier, we need to establish a well defined quiescent operating point or Q-point, so it must be biased to operate in its saturation region. The Q-point for the MOSFET is represented by the DC values I_D and V_{GS} that position the operating point centrally on the MOSFET's output characteristics curve. The DC characteristics and therefore Q-point are all functions of gate voltage V_G, supply voltage V_{DD} and load resistance R_D. As V_{GS} increases, the bias point moves up the curve as shown allowing a larger drain current to flow and V_{DS} decreases.

Likewise, as V_{GS} decreases, the bias point moves down the curve, resulting in a smaller drain current and increased V_{DS}. In order to establish a large output swing we must bias the transistor well above threshold level to ensure that the transistor stays in saturation over the full sinusoidal input cycle. However, there is a limit on the amount of gate bias and drain current we can use. To allow for maximum voltage swing of the output, the Q-point should be positioned approximately halfway between the supply voltage V_{DD} and the threshold voltage V_{th}.

The circuit combines the use of a source resistor R_S with a gate bias voltage, V_G. The gate bias voltage V_G is derived from the supply voltage V_{DD} by means of the voltage divider network consisting of resistors R_1 and R_2 and is given by

$$V_G = \frac{V_{DD}R_2}{R_1 + R_2}$$

The source voltage V_S which depends upon R_S and drain current I_D is

$$V_S = I_D R_S$$

The gate terminal and the source terminal are both positive with respect to ground, and the gate-source voltage V_{GS} is the difference between V_G and V_S. Because V_G is larger than V_S, the MOSFET gate terminal is at a higher potential than the source. So, the gate is positive with respect to the source, and the gate-channel junctions are forward biased. The voltage drop across R_S increases if I_D increases. This produces a decrease in V_{GS}, which drives I_D back toward its original level. A decrease in I_D causes an increase in V_{GS} which tends to increase I_D.

Assuming $\quad I_G = 0, V_{GS} = V_G - I_D R_S$

When $\quad I_D = 0, V_{GS} = V_G$

Circuit Diagram

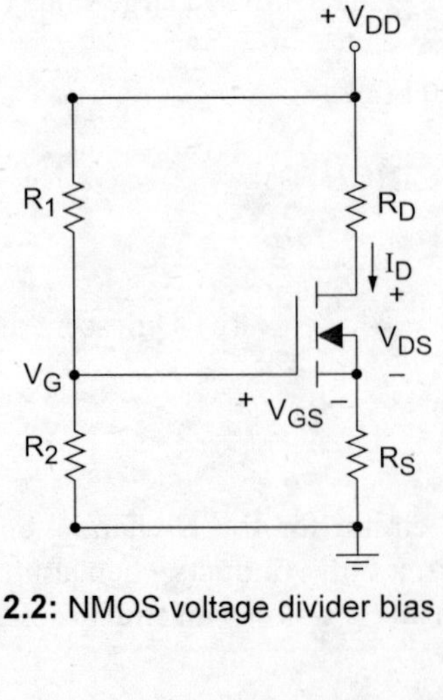

Fig. 2.2: NMOS voltage divider bias circuit

Design Equations

$I_D = \left(\frac{1}{2}\right)k_n(V_{GS} - V_t)^2$, where $k_n = \mu_n C_{ox}\frac{W}{L}$ is the technology parameter.

$V_{DS} \approx \frac{V_{DD}}{2}$ and $V_{RS} \approx 0.1V_{DD}$ for the MOSFET operating in saturation region.

$$V_{RD} = V_{DD} - V_{DS} - V_{RS}$$

The voltage at the gate terminal

$$V_G = \frac{V_{DD}R_2}{R_1 + R_2} = V_{GS} + V_{RS}$$

$$\frac{R_2}{R_1 + R_2} = \frac{V_G}{V_{DD}}$$

Design Steps

$$I_D = 30 \times 10^{-3} = \left(\frac{1}{2}\right)(180 \times 10^{-3})\,(V_{GS} - 2.1)^2$$

$$V_{GS} = 2.677 \text{ V or } 1.523 \text{ V}$$

Since, 1.523 V is $< V_t = 2.1$ V, $V_{GSQ} = 2.677$ V

Choosing $V_{DSQ} = \frac{V_{DD}}{2} = 10$ V and $V_{RS} = 0.1V_{DD} = 2$ V, $V_{RD} = 8$ V

$$R_D = 267\ \Omega$$

$$R_S = 66.67\ \Omega$$

$$V_G = 4.677 \text{ V}$$

$$\frac{R_2}{R_1 + R_2} = \frac{V_G}{V_{DD}} = 0.234$$

$$\frac{R_1}{R_2} = 3.274$$

Let $R_2 = 1.0$ MΩ. Then $R_1 = 3.274$ MΩ

2.3 COMMON SOURCE AMPLIFIER

Design Specifications

(*i*) Voltage gain $A_v = -10$ (min)

(*ii*) Input resistance $R_i \geq 400$ kΩ

(*iii*) Output resistance $R_o \leq 5$ kΩ

(*iv*) Peak-to-peak output voltage swing $V_{opp} = 6$ V

(*v*) Lower cut-off frequency $f_L < 1$ kHz

(*vi*) Supply voltage $V_{DD} = 20$ V

(*vii*) Drain current $I_D = 2$ mA.

Assume $\lambda = 0$ for the 2N7000 MOSFET.

Theory

A common source amplifier using NMOS is shown in Fig. 2.3. The source resistance R_S is split into two, R_{S1} and R_{S2}. R_{S2} is shunted by a large capacitance C_S. Both R_{S1} and R_{S2} set the DC bias point, while R_{S2} is effectively shorted for a small signal and gives the desired small-signal voltage gain. Assuming that the coupling capacitors C_1 and C_2, and the bypass capacitor C_S are large, they behave as short-circuited at the frequency of operation. Load resistance R_L is considered external to the amplifier and is not included.

The input resistance R_i of the amplifier $= \dfrac{v_g}{i_g} = R_1 \,||\, R_2 = R_G$

The output resistance R_o can be obtained by setting V_S equal to zero and then applying a test voltage v_x at the output side and let i_x is the corresponding test current.

$$R_o = \frac{v_x}{i_x} = [r_o + (1 + g_m r_o)R_{S1}] \,||\, R_D$$

where r_o is the NMOS output resistance and g_m is its transconductance.

The open-circuit voltage gain

$$A_{vo} = \frac{v_o}{v_g} = -\frac{g_m r_o R_D}{R_D + r_o + (1 + g_m r_o)R_{S1}}$$

The maximum open-circuit voltage gain,

$$A_{vo(\max)} = -\frac{g_m r_o R_D}{R_D + r_o}$$

For $r_o >> R_D$, $\quad A_{vo(\max)} = -g_m R_D$

The voltage gain, assuming r_o is very large, is given as

$$A_v = \frac{v_O}{v_S} = -\frac{g_m R_D}{1 + g_m R_S}$$

Circuit Diagram

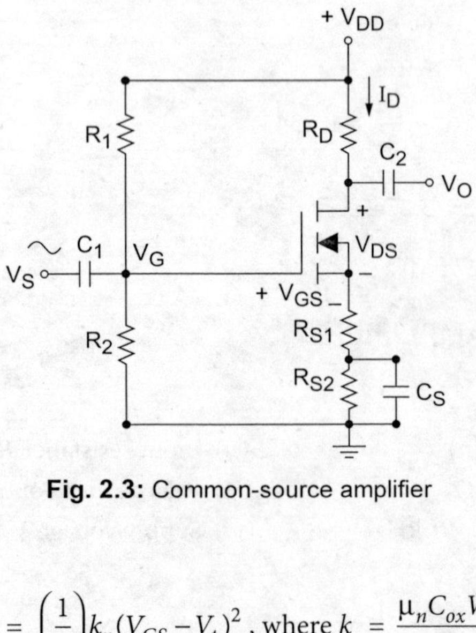

Fig. 2.3: Common-source amplifier

Design Equations

$$I_D = \left(\frac{1}{2}\right)k_n(V_{GS} - V_t)^2 \text{, where } k_n = \frac{\mu_n C_{ox} W}{L} \text{ is the technology parameter.}$$

$$V_{DS} \approx \frac{V_{DD}}{2} \text{ and } V_{RS} \approx 0.1\,V_{DD} \text{ for the MOSFET operating in saturation region.}$$

$$V_{RD} = V_{DD} - V_{DS} - V_{RS}$$

The voltage at the gate terminal,

$$V_G = \frac{V_{DD}R_2}{R_1+R_2} = V_{GS} + V_{RS}$$

$$\frac{R_2}{R_1+R_2} = \frac{V_G}{V_{DD}}$$

Input resistance $R_i = R_1 \,||\, R_2$

MOSFET transconductance $g_m = \dfrac{2I_D}{V_{GS}-V_t}$

MOSFET output resistance $r_o = \dfrac{1}{\lambda I_D}$

Voltage gain $A_v = -\dfrac{g_m R_D}{1+g_m R_S}$ when r_o is very large.

The output resistance of the amplifier $R_{out} = R_D$, neglecting r_o.

The input coupling capacitor C_1 can be found from

$$f_L = \frac{1}{2\pi(R_{sig} + R_1 \,||\, R_2)C_1}$$

where f_L is lower cut-off frequency and R_{sig} is signal source resistance.

The output coupling capacitor C_2 can be found from

$$f_L = \frac{1}{2\pi R_D C_2}, \text{ neglecting } r_o$$

The bypass capacitor C_S can be found from

$$f_L = \frac{1}{2\pi\left\{R_{S2} \,||\left(R_{S1} + \dfrac{1}{g_m}\right)\right\}C_S}$$

when R_S is split into R_{S1} and R_{S2}, with R_{S2} bypassed by C_S.

Design Steps

$$I_D = 2 = \left(\frac{1}{2}\right) 180\,(V_{GS} - 2.1)^2 \rightarrow V_{GS} = 2.25 \text{ or } 1.951 \text{ V}$$

Since $1.951 \text{ V} < V_t = 2.1 \text{ V}, V_{GS} = 2.25 \text{ V}$

Choosing, $V_{DS} = \dfrac{V_{DD}}{2} = 10 \text{ V}$ and $V_{RS} = 0.1V_{DD} = 2 \text{ V}, V_{RD} = 8 \text{ V}$

$$R_D = \frac{V_{RD}}{I_D} = 4 \text{ k}\Omega \quad \text{and} \quad R_S = \frac{V_{RS}}{I_D} = 1 \text{ k}\Omega$$

We need peak-to-peak output voltage swing, $V_{opp} = 6$ V. $V_{DS} > (V_{GS} - V_t)$

i.e., $10 \text{ V} > (2.25 - 2) = 0.25 \text{ V}$ ... Satisfied.

$$V_G = 4.25 \text{ V}$$

$$\frac{R_2}{R_1+R_2} = 0.2125$$

Given, $R_i \geq 400\ \text{k}\Omega.$

Choosing $R_i = 1\ \text{M}\Omega$

We have $0.2125R_1 = 1\ \text{M}\Omega$

$R_1 = 4.706\ \text{M}\Omega$ and $R_2 = 1.27\ \text{M}\Omega$

Given, $A_v = -10$ (min)

When C_S is not present,

$$A_v = -\frac{g_m R_D}{1+g_m R_S}, \text{ as } r_o = \frac{1}{\lambda I_D} = \frac{1}{0} = \infty$$

$$g_m = 26.67\ \text{mA/V}$$

$$|A_v| = 3.855, \quad \textit{i.e.}, < 10 \text{ required.}$$

In order to improve the gain, split R_S into R_{S1} and R_{S2}, with R_{S2} bypassed by C_S, as shown in Fig. 2.3.

Then, $$A_v = -\frac{g_m R_D}{1+g_m R_{S1}}$$

$$R_{S1} = 363\ \Omega$$

$$R_{S2} = R_S - R_{S1} = 1000 - 363 = 637\ \Omega$$

$$R_{out} = R_D = 4\ \text{k}\Omega$$

Neglecting R_{sig}, $$f_L = \frac{1}{2\pi(R_1 \,||\, R_2)C_1}$$

Choosing $f_L = 100$ Hz, $C_1 = 1.5915$ nF

$$f_L = \frac{1}{2\pi R_D C_2}$$

$$100 = \frac{1}{2\pi(4\ \text{k}\Omega)C_2} \quad \text{or} \quad C_2 = 0.398\ \mu\text{F}$$

$$f_L = \frac{1}{2\pi\left\{R_{S2} \,||\left(R_{S1} + \dfrac{1}{g_m}\right)\right\}C_S}$$

$$C_S = 6.5\ \mu\text{F}$$

All resistors are of 1/4 W and all capacitors are of 25 V rating.

2.4 SOURCE-FOLLOWER

Design Specifications

(*i*) Input resistance $R_i = 250\ \text{k}\Omega$ (*ii*) Supply voltage $V_{DD} = 10$ V

(*iii*) Quiescent drain current $I_{DQ} = 2$ mA (*iv*) Signal source resistance $R_{sig} = 3\ \text{k}\Omega$.

Assume $\lambda = 0$ for the 2N7000 MOSFET. Determine the small-signal parameters of MOSFET and the small-signal voltage gain.

Theory

A general common-drain or source-follower configuration is shown in Fig. 2.4. A common-drain amplifier has a very high input resistance and draws a very small gate current. It also offers a low output resistance and can be used as a buffer stage between a low resistance load and a signal source that can supply only a very small current. This configuration has a voltage gain approaching unity and is known as a source follower. In the circuit, the resistors R_1 and R_2 forms the biasing circuit that fixes V_G. R_{sig} is the input signal source resistance and C_1 is the input coupling capacitance. As seen in the figure, the output signal is taken off the source with respect to ground and the drain is connected directly to V_{DD}. Since V_{DD} becomes signal ground in the A.C. equivalent circuit, we have the name common drain.

The input and output relationship can be derived if we assume v_G is the input gate voltage and v_O is the output voltage at the source terminal. The gate-to-source voltage v_{GS}, which controls the drain current is given by

$$v_{GS} = v_G - v_O$$

The corresponding drain current, which must also flow through the source resistance R_S is

$$i_D = k_n(v_{GS} - v_t)^2$$

Neglecting body effect, the output voltage is

$$v_O = g_m v_{gs}(R_S \| r_o)$$

where g_m is the transconductance and r_o is the output resistance of the MOSFET.

The small-signal voltage gain is given by

$$A_v = \frac{v_O}{v_S} = \left[\frac{g_m(R_S \| R_o)}{1+g_m(R_S \| r_o)}\right] \cdot \frac{R_i}{R_i + R_{sig}}$$

where $R_i = R_1 \| R_2$ is the input resistance of the amplifier.

Circuit Diagram

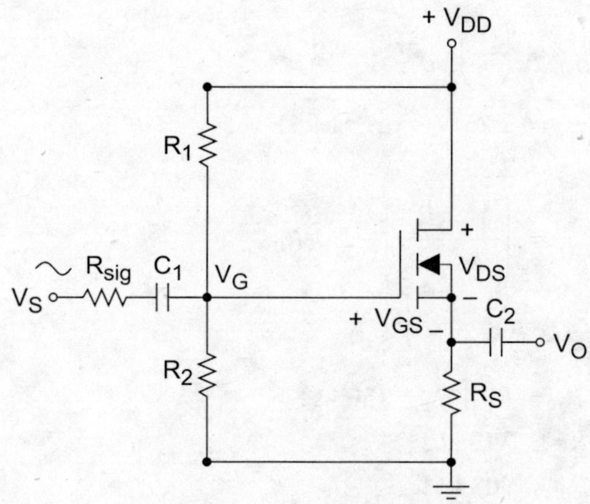

Fig. 2.4: Source-follower circuit

Design Equations

$$I_D = \left(\frac{1}{2}\right) k_n(V_{GS} - V_t)^2, \text{ where } k_n = \frac{\mu_n C_{ox} W}{L} \text{ is the technology parameter.}$$

$$V_{DS} \approx \frac{V_{DD}}{2} \text{ and } V_{RS} \approx 0.1 V_{DD} \text{ for the MOSFET operating in saturation region.}$$

Applying KVL around the drain-source loop, we have

$$V_{DD} = V_{DSQ} + I_{DQ} R_S$$

$$V_{GSQ} = V_{GQ} - V_{SQ} = \left[\frac{V_{DD}R_2}{R_1 + R_2}\right] - I_{DQ}R_S$$

The voltage at the gate terminal

$$V_G = \frac{V_{DD}R_2}{R_1 + R_2} = V_{GS} + V_{RS}$$

$$\frac{R_2}{R_1 + R_2} = \frac{V_G}{V_{DD}}$$

Input resistance $R_i = R_1 \,||\, R_2$

MOSFET transconductance $g_m = \dfrac{2I_D}{V_{GS} - V_t}$

MOSFET output resistance, $r_o = \dfrac{1}{\lambda I_D}$

The small-signal voltage gain, $A_v = \dfrac{V_o}{V_S} = \left[\dfrac{g_m R_S}{1 + g_m R_S}\right]\left[\dfrac{R_i}{R_i + R_{sig}}\right]$

Design Steps:

Choosing $V_{DSQ} = \dfrac{V_{DD}}{2} = 5\text{ V}$

$$R_S = 2.5\text{ k}\Omega$$

$$I_{DQ} = \left(\frac{1}{2}\right)k_n(V_{GSQ} - V_t)^2$$

$$V_{GSQ} = 2.25\text{ V} \quad \text{or} \quad 1.951\text{ V}$$

Since, $1.951\text{ V} < V_t$, $V_{GSQ} = 2.25\text{ V}$

$$V_{GSQ} = \left[\frac{V_{DD}R_2}{R_1 + R_2}\right] - I_{DQ}R_S$$

$$\frac{R_2}{R_1 + R_2} = 0.725$$

Given, $R_i = R_1 \,||\, R_2 = 250\text{ k}\Omega$

$$R_1 = 345\text{ k}\Omega$$

$$R_2 = 908\text{ k}\Omega$$

$$g_m = 26.67\text{ mA/V}$$

$$r_o = \infty$$

$$A_V = 0.9735$$

Chose $C_1 = C_2 = 1\ \mu\text{F, 15 V}$

2.5 CASCODE AMPLIFIER

Design Specifications

(*i*) DC gain = 70 dB (*ii*) $R_L = R_{\text{out}}$

(*iii*) $C_L = 0.1$ pF.

Neglect body effect. Determine the unity gain and 3-dB frequencies. For the 2N7000 transistors, $\lambda = 0$, $V_A = 50$ V, and $C_{gd} = 1$ pF.

Theory

Figure 2.5(*a*) shows the MOS cascode amplifier. Here transistor Q_1 is connected in the common-source configuration and provides its output to the input terminal, *i.e.*, source of transistor Q_2. Transistor Q_2 has a constant d.c. voltage V_{BIAS}, applied to its gate. Thus, the signal voltage at the gate of Q_2 is zero, and Q_2 is operating as a *CG* amplifier with a constant current load, *I*. Obviously both Q_1 and Q_2 will be operating at d.c. drain currents equal to *I*. The feedback in the overall circuit that incorporates the cascode amplifier establishes an appropriate d.c. voltage at the gate of Q_1 so that its drain current is equal to *I*. Also, the value of V_{BIAS} has to be chosen so that both Q_1 and Q_2 operate in the saturation region at all times.

In response to the input signal voltage V_S, the common-source transistor Q_1 conducts a current signal $g_{m1}V_i$ in its drain terminal and feeds it to the source terminal of the common-gate transistor Q_2, called the cascode transistor. Transistor Q_2 passes the signal current $g_{m1}V_S$ on to its drain, where it is supplied to a load resistance R_L at a very high output resistance R_{out}. The cascode transistor Q_2 acts in effect as a buffer, presenting a low input resistance to the drain of Q_1 and providing a high resistance at the amplifier output.

Figure 2.5(*b*) shows the cascode circuit prepared for small-signal analysis and with a resistance R_L shown at the output. R_L is assumed to include the output resistance of current source I as well as an actual load resistance, if any. The *CS* transistor Q_1 provides the cascode amplifier with an infinite input resistance. Also, at the drain of Q_1 looking "downward," the output resistance of the *CS* transistor Q_1 is r_{01}. Looking "upward," the input resistance of the *CG* transistor Q_2 is

$$R_{in2} = \frac{1}{g_{m2} + g_{mb2}} + \frac{R_L}{A_{vo2}}$$

where $A_{vo2} = 1 + (g_{m2} + g_{mb2})r_{o2}$

The output resistance $R_{out} = r_{o2} + A_{vo2}r_{o1}$

The open-circuit voltage gain A_{vo} of the cascode amplifier for equal intrinsic gains of *CS* and *CG* stages $= -(g_m r_o)^2$

Circuit Diagrams

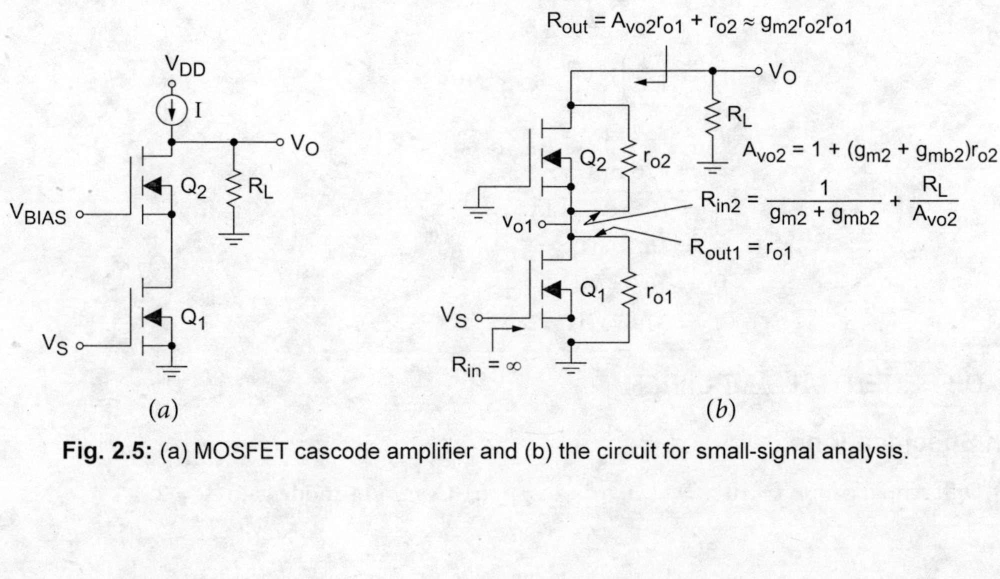

Fig. 2.5: (a) MOSFET cascode amplifier and (b) the circuit for small-signal analysis.

Design Equations

$$I_D = \left(\frac{1}{2}\right)k_n V_{ov}^2, \text{ where } V_{ov} = V_{GS} - V_t$$

$$A_v = \frac{A_{vo}R_L}{R_L + R_{out}}, \text{ where } A_{vo} \text{ is open-circuit voltage gain, given by}$$

$$A_{vo} = (1 + g_{m2}r_{o2})g_{m1}r_{o1} = g_m^2 r_o^2 = \left[\left(\frac{2I_D}{V_{ov}}\right)\cdot\left(\frac{V_A}{I_D}\right)\right]^2 = \left(\frac{2V_A}{V_{ov}}\right)^2$$

where V_A = Early voltage

$$r_o = \frac{V_A}{I_D}$$

$$g_m = \frac{2I_D}{V_{ov}}$$

For R_{sig} small, the time constant, $\tau_H = (R_L \,||\, R_{out})(C_{gd} + C_L)$

$$R_{out} = A_{vo2}r_{o1} + r_{o2} = (1 + g_m r_o)r_o + r_o$$

The upper 3-dB frequency, $f_H = \dfrac{1}{2\pi\tau_H}$

The unity-gain frequency, $f_t = |A_v| f_H$

Design Steps

$$A_v = 70 \text{ dB} = 3162$$

$$A_v = \frac{A_{vo}R_L}{R_L + R_{out}} = \frac{A_{vo}}{2}$$

$$A_{vo} = \left(\frac{2V_A}{V_{ov}}\right)^2$$

$$3162 = \frac{(2 \times 50/V_{ov})^2}{2} \quad \text{or} \quad V_{ov} = 1.257 \text{ V}$$

$$I_D = \left(\frac{1}{2}\right)k_n V_{ov}^2 = 142 \text{ mA}$$

$$r_o = 352\ \Omega$$

$$g_m = 226 \text{ mA/V}$$

$$R_{out} = R_L = 28.71 \text{ k}\Omega$$

$$\tau_H = 15.79 \text{ ns}$$

$$f_H = 10.08 \text{ MHz}$$

$$f_t = 31.87 \text{ GHz}$$

2.6 DIFFERENTIAL AMPLIFIER

Design Specifications

(*i*) Differential gain $A_d = 10$

(*ii*) Common-mode gain $A_c = 2$

(*iii*) Supply voltage $V_{DD} = 5$ V.

The dc bias current of 1 mA is provided by the resistor R_{SS} and the dc common-mode voltage is V_{CM}.

Assume $\lambda = 0$ for the 2N7000 NMOS transistors. Find the required values of R_D, R_{SS}, V_{CM}, and the dc voltage at the drains.

Theory

Figure 2.6 shows the basic MOS differential-pair configuration. It consists of two matched transistors Q_1 and Q_2 whose sources are joined together and biased by a constant-current source I. Assume that the current source is ideal and that it has infinite output resistance. Each drain is connected to the positive supply through a resistance R_D. It is essential that the MOSFETs not enter the triode region of operation.

First, assume that the two gate terminals are joined together and connected to a voltage v_{CM}, called the common-mode voltage. That is, as shown in figure, $v_{G1} = v_{G2} = v_{CM}$. Since Q_1 and Q_2 are matched, it follows from symmetry that the current I will divide equally between the two transistors. Thus, $i_{D1} = i_{D2} = I/2$, and the voltage at the sources v_S, will be

$$v_S = v_{CM} - v_{GS}$$

where v_{GS} is the gate-to-source voltage corresponding to a drain current of $I/2$. The voltage at each drain will be

$$v_{o1} = v_{o2} = V_{DD} - \left(\frac{I}{2}\right)R_D$$

Thus, the difference in voltage between the two drains will be zero.

If R_{SS} is the resistance of the current source I, then the common-mode gain of the circuit shown is given by

$$A_c = \frac{v_{o1}}{v_{CM}} = \frac{-R_D}{1/g_m + 2R_{SS}}$$

With differential input voltage, $v_{id} = v_{GS1} - v_{GS2}$, the voltages at the gate terminals of Q_1 and Q_2 are

$$v_{G1} = \frac{v_{CM} + v_{id}}{2}$$

$$v_{G2} = \frac{v_{CM} - v_{id}}{2}$$

The voltages at the drain terminals are

$$v_{o1} = -g_m\left(\frac{v_{id}}{2}\right)R_D$$

$$v_{o2} = +g_m\left(\frac{v_{id}}{2}\right)R_D$$

Therefore, the single-ended and differential gains are

$$\frac{v_{o1}}{v_{id}} = -\left(\frac{1}{2}\right)g_m R_D$$

$$\frac{v_{o2}}{v_{id}} = +\left(\frac{1}{2}\right)g_m R_D$$

$$A_d = \frac{v_{o2} - v_{o1}}{v_{id}} = g_m R_D$$

Circuit Diagram

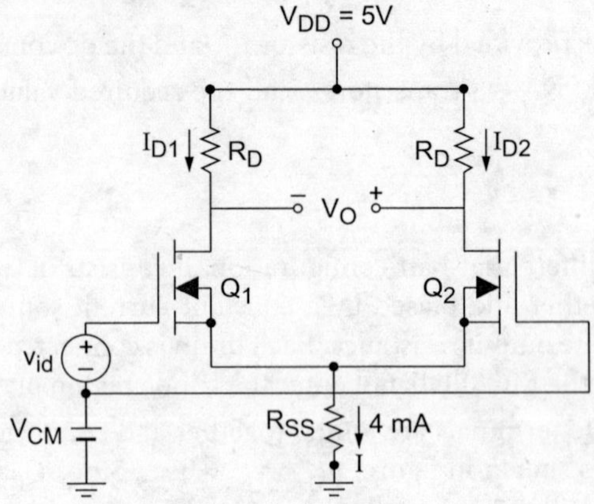

Fig. 2.6: MOSFET differential amplifier

Design Equations

$$I_{D1} = I_{D2} = \frac{I}{2}$$

$$I_D = \left(\frac{1}{2}\right) k_n V_{ov}^2 \text{, where } V_{ov} = V_{GS} - V_t$$

$$g_m = \frac{I}{V_{ov}}$$

The differential gain, $A_d = g_m R_D$

Voltages at the drains are $V_{D1} = V_{D2} = V_{DD} - I_D R_D$

The common-mode gain, $A_c = \frac{V_{o1}}{V_{CM}} = \frac{-R_D}{1/g_m + 2R_{SS}}$

The input common-mode voltage

$$V_{CM} = V_{GS} + V_S$$

Design Steps

$$I_{D1} = I_{D2} = \frac{I}{2} = 2 \text{ mA}$$

$$V_{ov} = 0.149 \text{ V}$$

$$V_{GS} = 2.249 \text{ V}$$

$$g_m = 26.85 \text{ mA/v}$$

$$R_D = \frac{A_d}{g_m} = 372.44\ \Omega$$

Voltages at the drains are $V_{D1} = V_{D2} = 4.255$ V

$$|A_c| = 2 \text{ gives } R_{SS} = 74.5\ \Omega$$

Voltage at the source terminals $V_S = IR_{SS} = 0.298$ V

The input common-mode voltage $V_{CM} = 2.547$ V

2.7 SIMPLE CURRENT MIRROR

Design Specifications

(*i*) Reference current $I_{REF} = I_O = 10$ mA (*ii*) Supply voltage $V_{DD} = 10$ V.

For the NMOS transistor, assume $\lambda = 0$. Find I_o as a function of V_{out} and hence r_{out} of the circuit.

Theory

Figure 2.7 shows the circuit of a simple MOS current mirror. The heart of the circuit is transistor Q_1, the drain of which is shorted to its gate, thereby forcing it to operate in the saturation mode with

$$I_{D1} = \left(\frac{k_n'}{2}\right)\left(\frac{W}{L}\right)_1 (V_{GS} - V_{tn})^2$$

neglecting channel-length modulation. The drain current of Q_1 is supplied by V_{DD} through resistor R. Since the gate currents are zero,

$$I_{D1} = I_{REF} = \frac{(V_{DD} - V_{GS})}{R}$$

where the current through R is considered to be the reference current of the current source and is denoted I_{REF}.

Now consider transistor Q_2. It has the same V_{GS} as Q_1; thus, if we assume that it is operating in saturation, its drain current, which is the output current I_O of the current mirror, neglecting channel-length modulation, will be

$$I_O = I_{D2} = \left(\frac{k_n'}{2}\right)\left(\frac{W}{L}\right)_2 (V_{GS} - V_{tn})^2$$

Now, the output current I_O can be related to the reference current I_{REF} as

$$\frac{I_O}{I_{REF}} = \left(\frac{W}{L}\right)_2 \bigg/ \left(\frac{W}{L}\right)_1$$

The special connection of Q_1 and Q_2 provides an output current I_O that is related to the reference current I_{REF} by the ratio of the aspect ratios of the transistors. In other words, the relationship between I_O and I_{REF} is solely determined by the geometries of the transistors. In the special case of identical transistors, $I_O = I_{REF}$, and the circuit simply replicates or mirrors the reference current in the output terminal. This has given the circuit composed of Q_1 and Q_2 the name current mirror, a name that is used irrespective of the ratio of device dimensions.

Circuit Diagram

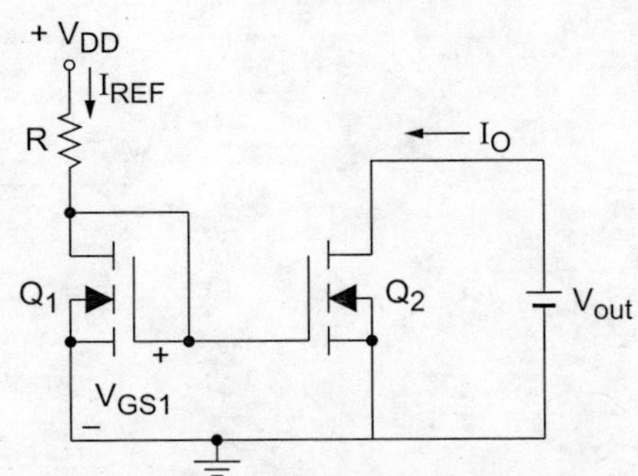

Fig. 2.7: MOSFET simple current mirror

Design Equations

$$I_{D1} = I_{REF} = \left(\frac{1}{2}\right)k_n(V_{GS1} - V_t)^2$$

$$R = \frac{V_{DD} - V_{GS1}}{I_{REF}}$$

Design Steps

For Q_1,

$$10 = \left(\frac{1}{2}\right)180(V_{GS1} - 2.1)^2$$

$$V_{GS1} = 2.433 \text{ V} \quad \text{or} \quad 1.767 \text{ V}$$

Since, $1.767 \text{ V} < V_t = 2.1 \text{ V}, \quad V_{GS1} = 2.433 \text{ V}$

$$R = 757\ \Omega$$

❑❑❑

3

CHAPTER

BJT CIRCUITS

3. BJT CIRCUITS

The bipolar junction transistor or BJT, as the name suggests, makes use of both electrons and holes for conduction. It consists of three terminals, namely, emitter (E), base (B), and collector (C) and two junctions, emitter-base or emitter junction and collector-base or collector junction. The emitter is highly doped, base is lightly doped, whereas the doping is moderate in the collector region. To collect more current at the collector terminal, the base region is made thin. The area of collector is made larger to handle more power. BJTs are of two types, NPN and PNP. In NPN transistors, electrons are the majority carriers and holes are the minority carriers. BJT is a current controlled device where small current at the base terminal is used to control large current at other terminals.

BJTs are operated in different modes. In forward active mode operation, emitter junction is forward biased and collector junction is reverse biased. When emitter junction is reverse biased and collector junction is forward biased, the BJT is said to be operated in reverse active mode. The saturation region operation results when both the junctions are forward biased. The device is in cut-off mode when both the junctions are reverse biased. BJTs are used as switches and for amplifying the signals.

In this chapter, design of BJT switch circuit, series and shunt voltage regulators, voltage divider bias circuit, common emitter amplifier, emitter follower, cascode amplifier, differential amplifier, simple current source, feedback amplifiers, oscillators, single-tuned amplifier, class *A*, class *B*, and class *AB* power amplifiers, for given design specifications, is given.

3.1 SWITCH CIRCUIT

Theory

The schematic to operate the BJT as a switch is shown in Fig. 3.1. The common-emitter configuration shown is to be operated in cut off and saturation modes for this purpose. The circuit shown consists of a manual switch S, base resistor R_B, collector resistor R_C and a LED to indicate the switching condition of BJT.

With the manual switch S is open, the base voltage is less than $V_{BE}(on)$ required for the transistor to conduct and BJT will be cut off; thus $i_B = 0$, $i_C = 0$, and $v_C = V_{CC}$. In this state, node C is disconnected from ground, the BJT switch is in the open position and LED will not glow.

The base current with S closed is

$$i_B = \frac{V_{CC} - V_{BE}(on)}{R_B}$$

and the corresponding collector $i_C = \beta i_B$

For sufficiently large i_B, *i.e.*, $i_B > i_B$(min) required to drive the transistor into saturation, the BJT conducts heavily and the LED glows.

Circuit Diagram

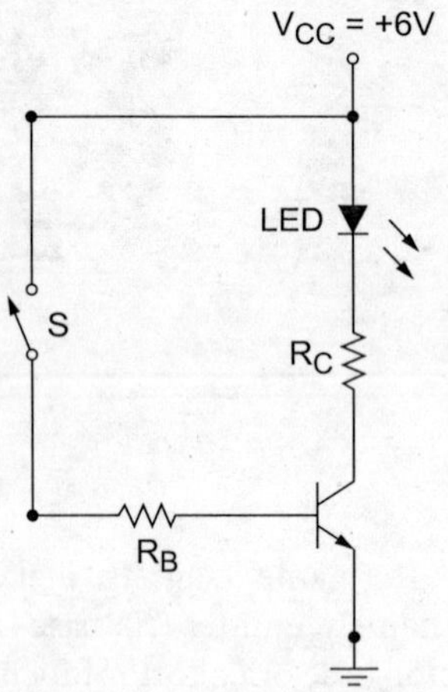

Fig. 3.1: BJT switch circuit

Design Equations

$$R_C = \frac{V_{CC} - V_{CEsat}}{I_C} \quad \text{and} \quad I_B = \frac{I_C}{\beta}$$

$$R_B = \frac{V_{CC} - V_{BE}}{I_B}$$

Design Steps

Assume $V_{CC} = +6$ V

Choosing Si NPN transistor 2N2222, we have $I_{Cmax} = 800$ mA, $\beta_{min} = 100$, $\beta_{typ} = 175$, and $V_{CEsat} = 0.4$ V. Assume $V_{BE} = 0.7$ V.

Choose IR323 LED.

To limit the current through LED, *i.e.*, I_C to 20 mA when the transistor is operating in saturation,

$$R_C = 280\ \Omega$$

$$I_B = 0.1143\ \text{mA}$$

$$R_B = 46.4\ \text{k}\Omega$$

3.2 SERIES VOLTAGE REGULATOR

Design Specifications

(*i*) Unregulated input voltage $V_i = 20$ V and $r_o = 10\ \Omega$

(*ii*) Regulated output voltage $V_o = 12$ V (*iii*) Maximum load current $I_{Lmax} = 100$ mA.

Use Si transistors SL100 (Q_1) and 2N2222A (Q_2). The parameters of SL100 are $I_C = 0.5$ A, total power dissipation, $P_{tot} = 0.8$ W, $\beta_{min} = 40$, and $\beta_{max} = 300$. For 2N2222A, $I_C = 0.6$ A, total power dissipation, $P_{tot} = 0.5$ W, $\beta_{min} = 40$, $\beta_{max} = 300$, $\beta_{typ} = 175$, and $r_\pi = 1$ kΩ.

Theory

Figure 3.2 shows the series voltage regulator circuit using transistors. The voltage regulation occurs as a large fraction of the increase in input voltage that appears across the pass transistor. If the input increases, the output must also increase, because it is this increase in output that acts to bias the pass transistor toward less current. In the circuit, Q_2 is the comparison amplifier and the reference voltage V_R is provided by the Zener diode D. Here a fraction of the output voltage $\beta' V_O$ is compared with the reference voltage V_R. The difference $(\beta' V_O - V_R)$ is amplified by Q_2. If the input voltage increases by ΔV_i, then V_O need increase only slightly, and yet Q_2 may cause a large current change in R_3. Thus, it is possible for almost all of ΔV_i to appear across R_3 and for V_O to remain essentially constant. The output DC voltage is given by

$$V_O = V_R + V_{BE2} + \frac{V_O R_1}{R_1 + R_2}$$

The input regulation factor S_V is given as

$$S_V = \frac{\Delta V_O}{\Delta V_i} = \frac{1}{G_m R_3}$$

where $G_m = \left[\frac{\beta_2 R_2}{R_1 + R_2}\right] \cdot \left[\frac{1}{(R_1 \| R_2) + r_{\pi 2} + (1+\beta_2) r_z}\right]$ and r_z is the dynamic resistance of the Zener diode.

Circuit Diagram

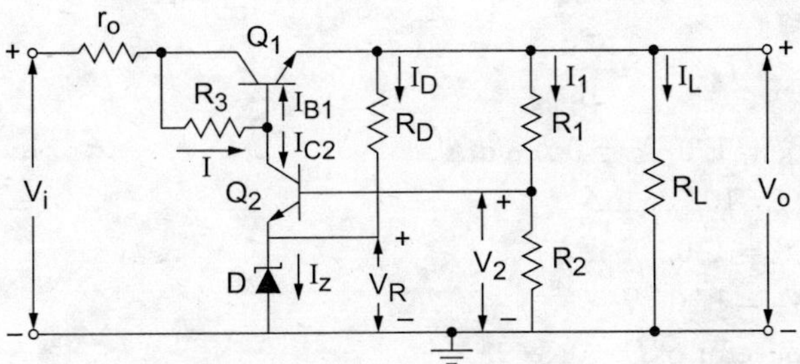

Fig. 3.2: Series voltage regulator

Design Equations

Reference voltage to be provided by Zener diode, $V_R \approx \frac{V_O}{2}$

From the circuit diagram,

$$I_Z = I_{E2} + I_D$$

$$R_D = \frac{V_o - V_R}{I_D} = \frac{V_o - V_Z}{I_D} \quad \text{and} \quad I_{B2} = \frac{I_{C2}}{\beta_2}$$

$$V_2 = V_{BE2} + V_R$$

$$R_1 = \frac{V_o - V_2}{I_1} \quad \text{and} \quad R_2 \approx \frac{V_2}{I_1}$$

$$I_{B1} \approx \frac{I_{E1}}{\beta_1} = \frac{I_L + I_1 + I_D}{\beta_1}$$

The current through R_3, $I = I_{B1} + I_{C2}$

The current through r_o, $I_{ro} = I_{C1} + I \approx I_{E1} + I$

The voltage V_{C1} at the collector of $Q_1 = V_i - V_{ro}$

$$R_3 = \frac{V_{C1} - (V_{BE1} + V_O)}{I}$$

$$R_L = \frac{V_O}{I_L}$$

Design Steps

Choose a Si reference diode with $V_R \approx V_O/2 = 6$ V. From the datasheets of Zener diodes, 1N753 can be chosen. The parameters of 1N753 are, nominal Zener voltage $V_Z = 6.2$ V, maximum Zener impedance $r_{Zmax} = 7\Omega$ at $I_Z = 20$ mA, power dissipation $P_{Zmax} = 500$ mW, and maximum reverse current $I_R = 0.1$ μA at $V_R = 1$ V.

Also, choose $I_{C2} \approx I_{E2} = 10$ mA and $I_D = 10$ mA, so that D operates at $I_Z = 20$ mA.

$$R_D = 580\ \Omega$$

$$I_{B2} = 57.14\ \mu\text{A}$$

Choose $I_1 >> I_{B2}$ as 10 mA

$$V_2 = 6.9\ \text{V}$$

$$R_1 = 510\ \Omega$$

$$R_2 \approx 690\ \Omega$$

$$I_{B1} \approx \frac{120\ \text{mA}}{\beta_1}$$

Taking typical value of β_1 as 170, $I_{B1} = 0.706$ mA

The current through R_3, $I = 10.706$ mA

The current through r_o, $I_{ro} = 130.706$ mA

Voltage drop across r_o, $V_{ro} = 1.307$ V

The voltage V_{C1} at the collector of $Q_1 = 18.693$ V

$$R_3 = 560\ \Omega$$

$$R_L = 120\ \Omega$$

The power rating of R_L, $P_{RL} = I_L^2 R_L = 1.2$ W

The power dissipation of Q_1, $P_{Q1} = V_{CE1} I_{C1} \approx V_{CE1} \cdot I_{E1} = (V_{C1} - V_O) I_{E1} = 0.8$ W

This satisfies the given specifications of SL100 transistor.

3.3 SHUNT VOLTAGE REGULATOR

Design Specifications

(*i*) Unregulated input voltage $V_i = 15$ V (*ii*) Regulated output voltage $V_o = 9$ V

(*iii*) Maximum load current $I_{Lmax} = 200$ mA.

Use Si transistor SL100, whose parameters are $I_C = 0.5$ A, total power dissipation, $P_{tot} = 0.8$ W, $\beta_{min} = 40$, and $\beta_{max} = 300$.

Theory

A simple shunt regulator circuit using BJT is shown in Fig. 3.3. Resistor R_S drops the unregulated voltage V_i by an amount that depends on the current supplied to the load R_L. The voltage across the load is set by the Zener diode D and base-emitter voltage V_{BE} of transistor Q. If the load resistance decreases, the drive current to the base of Q reduces. This lowers the collector current and increases the load current, thereby maintaining the output voltage across the load constant. The output voltage is given by,

$$V_O = V_Z + V_{BE}$$

Circuit Diagram

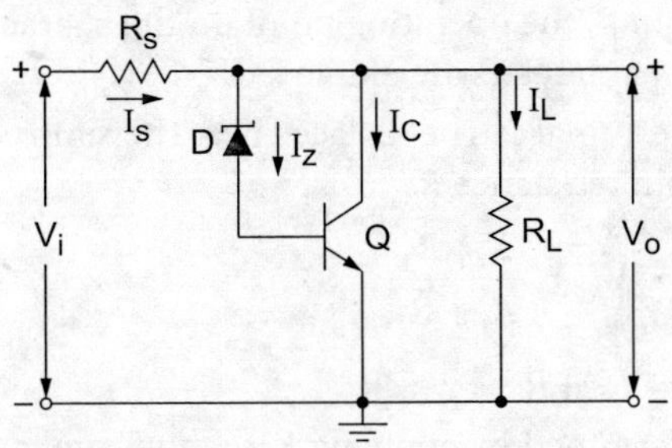

Fig. 3.3: Shunt voltage regulator

Design Equations

$$V_O = V_Z + V_{BE}$$

The current through R_S is $I_s = I_Z + I_C + I_L = I_B + I_C + I_L = I_B + \beta \cdot I_B + I_L$

The series resistance, $R_S = \dfrac{V_i - V_o}{I_s}$ and $R_L = \dfrac{V_O}{I_L}$

Design Steps

$$V_O = V_Z + V_{BE}$$

$$9 = V_Z + 0.7 \rightarrow V_Z \text{ required} = 8.3 \text{ V}$$

Choose Zener diode 1N4738A having V_z = 8.2 V, P_D = 1 W, I_{ZK} = 0.5 mA, I_{ZT} = 31 mA, Z_Z = 4.5 Ω, I_R = 10 µA at V_R = 6 V.

Assuming $I_Z = I_{ZK}$ when I_L is maximum, $I_s = 0.5 + \beta(0.5) + 200$ mA

Choosing typical value of β as 170, I_s = 285.5 mA

The series resistance, R_S = 21 Ω

Power dissipation of $R_S = P_{RS} = I_s^2 R_S$ = 1.71 W

$$R_L = 45\ \Omega$$

Power dissipation of $R_L = P_{RL} = \dfrac{V_O^2}{R_L} = 1.8$ W

Power dissipation of Q, $P_Q = V_{CE} I_C = V_o I_C = 0.765$ W

This satisfies the power rating of SL100 transistor.

3.4 VOLTAGE DIVIDER BIAS CIRCUIT

Design Specifications

(*i*) Quiescent collector current I_{CQ} = 10 mA (*ii*) Quiescent collector to emitter voltage V_{CEQ} = 5 V.

Use Si NPN transistor 2N2222 having $\beta_{min} = 100$, $\beta_{typ} = 175$, $I_S = 3.295 \times 10^{-14}$ A, and $V_A = 200$ V. Assume $V_{BE} = 0.7$ V.

Theory

The voltage divider bias circuit for BJT is shown in Fig. 3.4. It offers less sensitivity to changes in beta. Even though I_{BQ} changes with the change in β, the operating point on the characteristics defined by I_{CQ} and V_{CEQ} remain constant if the proper circuit parameters are employed.

The circuit to the left of the base terminal can be replaced by a Thevenin equivalent consisting of Thevenin's voltage V_{Th} in series with the Thevenin resistance R_{Th}.

$$V_{Th} = V_{R2} = \frac{V_{CC}R_2}{R_1 + R_2}$$

$$R_{Th} = R_1 \parallel R_2$$

The base current I_{BQ} can be determined by applying Kirchhoff's voltage law in the clockwise direction for the input loop.

$$V_{Th} - I_B R_{Th} - V_{BE} - I_E R_E = 0$$

Putting $I_E = (1 + \beta)I_B$, the base current $I_B = \dfrac{V_{Th} - V_{BE}}{R_{Th} + (1+\beta)R_E}$

$$I_C = \beta I_B$$

Applying KVL from V_{CC} to ground,

$$V_{CE} = V_{CC} - I_C R_C - I_E R_E = V_{CC} - I_C\left[R_C + \left(1 + \frac{1}{\beta}\right)R_E\right] \approx V_{CC} - I_C(R_C + R_E)$$

Circuit Diagram

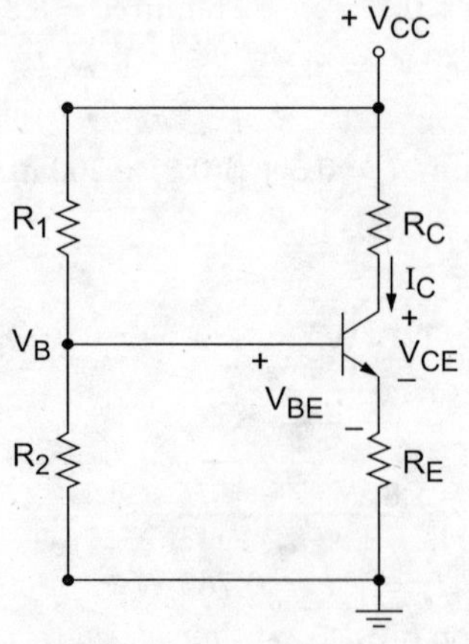

Fig. 3.4: BJT voltage divider bias circuit

Design Equations

$$V_{CC} \approx 2V_{CEQ}$$

$$I_B = \frac{I_C}{\beta}$$

$$\alpha = \frac{\beta}{1+\beta}$$

$$I_E = \frac{I_C}{\alpha}$$

Emitter voltage $V_E \approx 1/10^{th}$ of V_{CC}

$$R_E = \frac{V_E}{I_E} \quad \text{and} \quad R_C = \frac{V_{RC}}{I_C} = \frac{V_{CC} - V_{CE} - V_E}{I_C}$$

Thevenin's equivalent resistance at the base of transistor $R_{Th} = \frac{(1+\beta)R_E}{10}$

Thevenin voltage at the base terminal $V_{Th} = R_{Th}I_B + V_{BE} + V_E$

$$R_1 = \frac{R_{Th}V_{CC}}{V_{Th}} \quad \text{and} \quad R_2 = \frac{R_{Th}V_{CC}}{V_{CC} - V_{Th}}$$

Design Steps

Choose

$$V_{CC} = 10 \text{ V}$$
$$I_B = 0.06 \text{ mA}$$
$$\alpha = 0.994$$
$$I_E = 10.06 \text{ mA}$$

Taking emitter voltage as $1/10^{th}$ of V_{CC}, $V_E = 1$ V

$$R_E = 99.4\ \Omega$$
$$R_C = 400\ \Omega$$

Thevenin's equivalent resistance at the base of transistor, $R_{Th} = 1.75$ kΩ

Thevenin voltage at the base terminal, $V_{Th} = 1.805$ V

For the circuit to operate efficiently, it is assumed that the current through R_1 and R_2 is approximately equal and much larger than the base current.

$$R_1 = 9.7 \text{ k}\Omega$$
$$R_2 = 2.14 \text{ k}\Omega$$

3.5 COMMON EMITTER AMPLIFIER

Design Specifications

(*i*) Minimum voltage gain $\frac{V_o}{V_s} = 10$

(ii) Lower cut-off frequency $f_L = 100$ Hz

(*iii*) The *Q* point is to be located at $I_C = 3$ mA and $V_{CE} = 6$ V

(*iv*) Signal source resistance $R_s = 5$ kΩ

(*v*) Load resistance $R_L = 1$ kΩ.

The parameters of the Si NPN transistor 2N2222 are $\beta_{min} = 100$, $\beta_{typ} = 175$, and $V_A = 200$.

Theory

Figure 3.5 shows the circuit diagram of a *CE* amplifier and its small-signal equivalent. The *Q*-point to operate the BJT in the active region is established by the biasing circuit. An input voltage V_s is applied through coupling capacitor C_{C1} and output of the amplifier is coupled to the load R_L through another coupling capacitor C_{C2} as shown in figure. The emitter resistor R_E is shunted by bypass capacitor C_E. The capacitors C_{C1} and C_{C2} isolate the DC signals of the biasing circuit from the input signal V_s and the load resistance R_L, respectively. If the input signal were connected directly to the base without C_{C1}, the source resistance R_S would form a parallel circuit with R_2, and the base potential V_B would be disturbed. Similarly, the collector potential V_C would depend on R_L if C_{C2} were removed. Let us assume that the coupling and bypass capacitors have large values so that they are virtually shorted at the frequency of the input signal. With a sinusoidal input voltage $V_s = V_m \sin \omega t$, the base potential will be $v_B = V_B + v_b$. If the base current i_B swings between $I_B + i_b$(peak) and $I_B - i_b$(peak), the collector current i_C will swing between $I_C + i_c$(peak) and $I_C - i_c$(peak). The *C-E* voltage v_{CE} will vary accordingly from $V_{CE} + v_{ce}$(peak) to $V_{CE} - v_{ce}$(peak), and the collector voltage v_C will vary from $V_C - R_C(I_C + i_c\text{(peak)})$ to $V_C - R_C(I_C - i_c\text{(peak)})$. Since C_{C2} will block any DC signal, the output voltage will vary from $-(R_C \,||\, R_L)i_c$(max) to $(R_C \,||\, R_L)i_c$(min). The voltage gain of the amplifier circuit is given by

$$A_V = \frac{V_o}{V_s} = \frac{-g_m(r_o \,||\, R_C \,||\, R_L)\cdot(R_B \,||\, r_\pi)}{R_s + (R_B \,||\, r_\pi)}$$

where g_m is the transconductance and r_π and r_o are the input and output resistances of the transistor.

The input resistance, $R_i = R_B \,||\, r_\pi \approx r_\pi$, since $R_B >> r_\pi$

The output resistance, $R_o = r_o \,||\, R_C$

Circuit Diagram

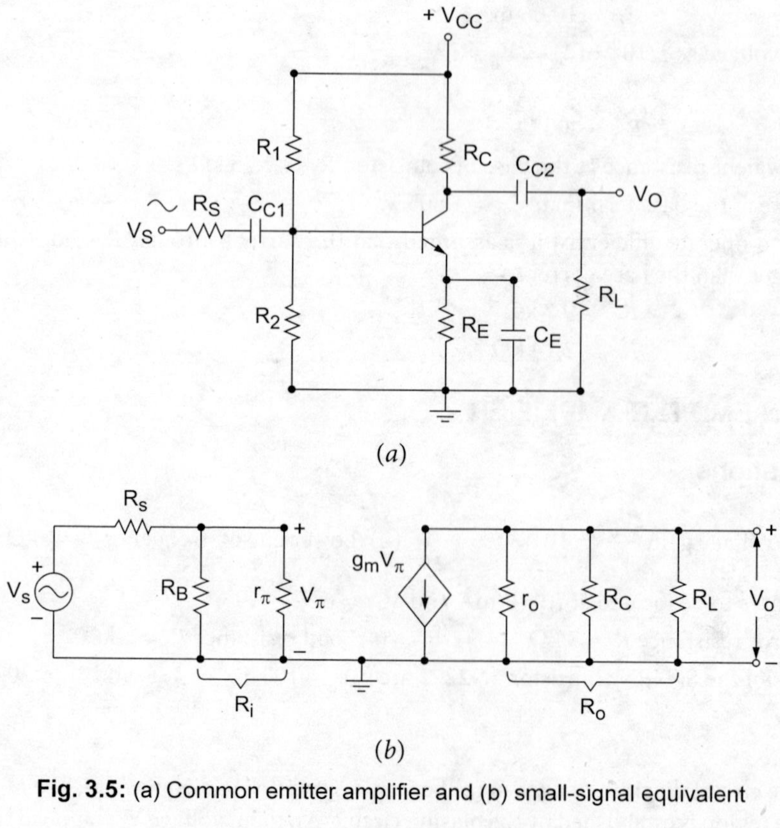

Fig. 3.5: (a) Common emitter amplifier and (b) small-signal equivalent

Design Equations

$$V_{CC} \approx 2\ V_{CEQ}$$

$$I_B = \frac{I_C}{\beta}$$

$$\alpha = \frac{\beta}{1+\beta}$$

$$I_E = \frac{I_C}{\alpha}$$

Emitter voltage $V_E \approx 1/10^{\text{th}}$ of V_{CC}

$$R_E = \frac{V_E}{I_E} \quad \text{and} \quad R_C = \frac{V_{RC}}{I_C} = \frac{V_{CC} - V_{CE} - V_E}{I_C}$$

Thevenin's equivalent resistance at the base of transistor, $R_{Th} = R_B = \dfrac{(1+\beta)R_E}{10}$

Thevenin voltage at the base terminal, $V_{Th} = R_{Th}I_B + V_{BE} + V_E$

$$R_1 = \frac{R_{Th}V_{CC}}{V_{Th}}$$

$$R_2 = \frac{R_{Th}V_{CC}}{V_{CC} - V_{Th}}$$

$$A_V = \frac{V_o}{V_s} = \left(\frac{V_o}{V_\pi}\right)\cdot\left(\frac{V_\pi}{V_s}\right)$$

$$\frac{V_o}{V_\pi} = -\,g_m R = -\,g_m(r_o \,||\, R_C \,||\, R_L)$$

$$g_m = \frac{I_C}{V_T}$$

$$r_\pi = \frac{\beta}{g_m}$$

$$r_o = \frac{V_A + V_{CE}}{I_C} \approx \frac{V_A}{I_C}$$

$$\frac{V_\pi}{V_s} = \frac{R_i}{R_s + R_i} \quad \text{and} \quad R_i = R_B \,||\, r_\pi$$

The input coupling capacitor $C_{C1} = \dfrac{1}{2\pi f_L(R_S + R_i)}$

The output coupling capacitor $C_{C2} = \dfrac{1}{2\pi f_L(R_C + R_L)}$

The bypass capacitor $C_E = \dfrac{1}{2\pi f_L R'_E}$

where $R'_E = \dfrac{R_E \,||\, [r_\pi + (R_S \,||\, R_B)]}{1+\beta}$

Design of Biasing Circuit

Choose,
$$V_{CC} = 12 \text{ V}$$
$$I_B = 0.017 \text{ mA}$$
$$\alpha = 0.994$$
$$I_E = 3.017 \text{ mA}$$

Taking emitter voltage as 1/10th of V_{CC}, $V_E = 1.2$ V
$$R_E = 398\ \Omega$$

Equivalent resistance at the base of transistor, $R_{Th} = R_B = 7$ kΩ

Thevenin voltage at the base terminal, $V_{Th} = 2.02$ V

For the circuit to operate efficiently, it is assumed that the current through R_1 and R_2 is approximately equal and much larger than the base current.
$$R_1 = 41.58 \text{ k}\Omega$$
$$R_2 = 8.42 \text{ k}\Omega$$

AC Design

$$A_V = -10$$
$$g_m = 116.3 \text{ mA/V}$$
$$r_\pi = 1.505 \text{ k}\Omega$$
$$r_o \approx 66.67 \text{ k}\Omega$$
$$R_i = 1.24 \text{ k}\Omega$$
$$\frac{V_\pi}{V_s} = 0.1987$$
$$-10 = 0.1987\,[-0.1163R]$$
$$R = 433\ \Omega$$
$$433 = 66.67 \text{ k}\Omega \,||\, R_C \,||\, 1 \text{ k}\Omega$$
$$R_C = 773\ \Omega$$

Given,
$$f_L = 100 \text{ Hz}$$
$$C_{C1} = 0.255\ \mu\text{F}$$
$$C_{C2} = 0.898\ \mu\text{F}$$
$$R'_E = 23.633\ \Omega$$
$$C_E = 67.34\ \mu\text{F}$$

3.6 EMITTER FOLLOWER

Design Specifications

(*i*) Voltage gain $V_o/V_i = 0.995$

(*ii*) Lower cut-off frequency $f_L = 100$ Hz

(*iii*) The Q point is to be located at $I_C = 10$ mA and $V_{CE} = 7.5$ V

(*iv*) Signal source resistance $R_S = 0.25$ kΩ

(*v*) Load resistance $R_L = 2$ kΩ.

The parameters of the Si NPN transistor 2N2222 are $\beta_{min} = 100$, $\beta_{typ} = 175$, and $V_A = 200$.

Theory

Figure 3.6 shows the circuit diagram of a *CC* amplifier or emitter follower. The Q-point to operate the BJT in the active region is established by the biasing circuit. An input voltage V_s is applied through coupling capacitor C_{C1} and output of the amplifier is coupled to the load R_L through another coupling capacitor C_{C2} as shown in figure. These capacitors C_{C1} and C_{C2} isolate the DC signals of the biasing circuit from the input signal V_s and the load resistance R_L, respectively. If the input signal were connected directly to the base without C_{C1}, the base potential V_B would be disturbed. Similarly, the emitter potential V_E would depend on R_L if C_{C2} were removed. Let us assume that the capacitors have large values so that they are virtually shorted at the frequency of the input signal. With a sinusoidal input voltage $V_s = V_m \sin \omega t$, the base potential will be $v_B = V_B + v_b$. The transistor output resistance r_o appears in parallel with R_L. The voltage gain of the emitter follower is given by

$$A_V = \frac{V_o}{V_i} = \frac{r_o \,||\, R_E \,||\, R_L}{r_e + (r_o \,||\, R_E \,||\, R_L)}$$

where V_i is the voltage at the base of transistor and r_e is the emitter resistance.

The input resistance of the amplifier circuit $R_i = R_B \,||\, (1 + \beta)[r_e + (r_o \,||\, R_E \,||\, R_L)]$

The output resistance $R_o = (r_o \,||\, R_E) \,||\left[\frac{r_e + (R_S \,||\, R_B)}{1+\beta}\right]$

Since r_o is much larger than R_E

$$R_o \approx R_E \,||\left[\frac{r_e + (R_S \,||\, R_B)}{1+\beta}\right]$$

The lower cut-off frequency due to C_{C1} is given by

$$f_{CC1} = \frac{1}{2\pi(R_s + R_i)C_{C1}}$$

The lower cut-off frequency due to C_{C2} is

$$f_{CC2} = \frac{1}{2\pi(R_L + R_o)C_{C2}}$$

Circuit Diagram

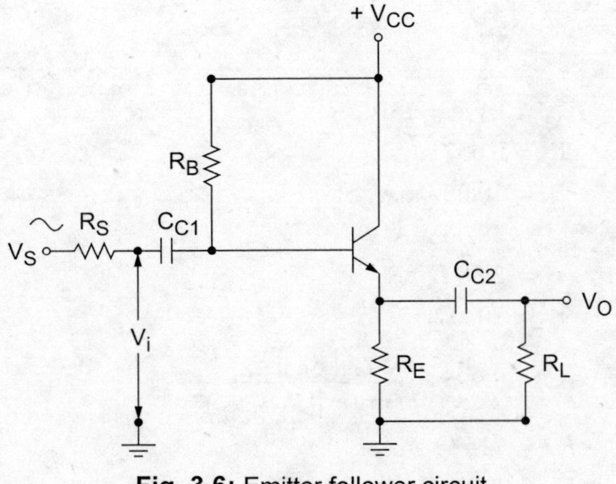

Fig. 3.6: Emitter follower circuit

Design Equations

$$V_{CC} \approx 2\,V_{CEQ}$$

$$I_B = \frac{I_C}{\beta}$$

$$\alpha = \frac{\beta}{1+\beta}$$

$$I_E = I_C + I_B$$

The voltage at the base, $V_B = V_{BE} + V_E$

$$R_B = \frac{V_{CC} - V_B}{I_B}$$

$$g_m = \frac{I_C}{V_T}$$

$$r_e = \frac{1}{g_m}$$

$$r_o \approx \frac{V_A}{I_C}$$

The voltage gain, $A_V = \dfrac{V_o}{V_i} = \dfrac{r_o \,||\, R_E \,||\, R_L}{r_e + (r_o \,||\, R_E \,||\, R_L)}$

The input resistance $R_i = R_B \,||\, (1+\beta)[r_e + (r_o \,||\, R_E \,||\, R_L)]$

The output resistance $R_o = (r_o \,||\, R_E) \,||\left[r_e + \dfrac{R_S \,||\, R_B}{1+\beta} \right]$

The lower cut-off frequency due to C_{C1}, $f_{CC1} = \dfrac{1}{2\pi(R_s + R_i)C_{C1}}$

The lower cut-off frequency due to C_{C2}, $f_{CC2} = \dfrac{1}{2\pi(R_L + R_O)C_{C2}}$

Design Steps

Choose,

$$V_{CC} = 2V_{CE} = 15\text{ V}$$

$$I_B = 0.057\text{ mA}$$

$$I_E = 10.057\text{ mA}$$

$$g_m = 387.6\text{ mA/V}$$

$$r_e = 2.58\ \Omega$$

$$r_o = 20\text{ k}\Omega$$

Given,

$$A_V = 0.995 = \frac{r_o \,||\, R_E \,||\, R_L}{r_e + (r_o \,||\, R_E \,||\, R_L)}$$

$$R_E = 715\ \Omega$$

$$V_E = 7.19\text{ V}$$

$$V_B = 7.89\text{ V}$$

$$R_B = 124.74\text{ k}\Omega$$

$$R_i = 52.54\text{ k}\Omega$$
$$R_o = 3.98\ \Omega$$

For
$$f_{CC1} = f_{CC2} = f_L,$$
$$C_{C1} = 30.15\text{ nF, 16 V}$$
$$C_{C2} = 0.794\ \mu\text{F, 16 V}$$

3.7 CASCODE AMPLIFIER

Design Specifications

(*i*) Voltage gain = – 50 (*ii*) $V_{CC} = 15$ V

(*iii*) $I_C = 3$ mA.

Use Si NPN transistors 2N2222 having $\beta_{min} = 100$, $\beta_{typ} = 175$, and $V_A = 200$.

Assume $V_{BE} = 0.7$ V.

Theory

A CE-CB, called as a cascode configuration is shown in Fig. 3.7. The input is fed to a common-emitter amplifier (Q_1), which drives a common-base amplifier (Q_2). As shown, the output signal current of Q_1 is the input signal of Q_2. One advantage of this circuit is that the output resistance looking into the collector of Q_2 is much larger than the output resistance of a simple common-emitter circuit. Another important advantage is in the frequency response.

If $\beta_1 = \beta_2 = \beta >> 1$, the base currents of Q_1 and Q_2 are negligible. Then, same current flows through the biasing resistors R_1, R_2, and R_3. The voltage at the base of Q_1 is

$$V_{B1} = \frac{V_{CC}R_3}{R_1 + R_2 + R_3}$$

and that at the base of Q_2 is $V_{B2} = \dfrac{V_{CC}(R_2 + R_3)}{R_1 + R_2 + R_3}$

The collector and emitter currents of transistors are

$$I_{E1} \approx I_{C1} = I_{E2} \approx I_{C2} = \frac{V_{B1} - V_{BE}}{R_E}$$

where $R_E = R_{E1} + R_{E2}$

The C-E voltages are

$$V_{CE1} = V_{B2} - V_{BE2} - (V_{B1} - V_{BE1}) = V_{B2} - V_{B1} \text{ (for } V_{BE1} = V_{BE2})$$
$$V_{CE2} = V_{CC} - R_C I_{C2} - (V_{B2} - V_{BE2})$$

The overall voltage gain, $A_V = \dfrac{V_o}{V_s} = A_{V1}A_{V2} = -g_{m1}g_{m2}\left[\dfrac{r_{\pi 2}}{1+\beta_2}\right]R_C \approx -g_{m1}R_C$

$$A_V = \left[\frac{R_{i2}}{r_{e1} + R_{E1}}\right]\cdot\left(\frac{R_C}{r_{e2}}\right) = \left[\frac{-r_{e2}}{r_{e1} + R_{E1}}\right]\cdot\left(\frac{R_C}{r_{e2}}\right) = \frac{-R_C}{r_{e1} + R_{E1}}$$

The input resistance of the circuit,

$$R_i = R_2 \,||\, R_3 \,||\, \beta(r_{e1} + R_{E1})$$

The output resistance $R_o = r_{o2} \,||\, R_C$

Circuit Diagram

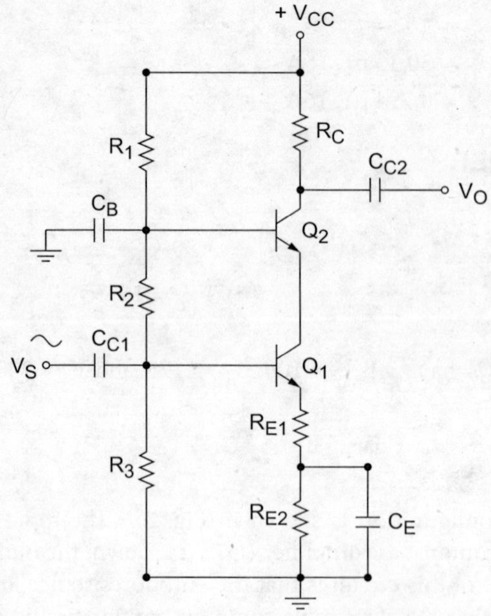

Fig. 3.7: BJT cascode amplifier

Design Equations

$$R_E = R_{E1} + R_{E2}$$

For better stability, choose the voltage at the emitter of Q_1 as $V_{E1} \approx 1/10^{th}$ of V_{CC}

$$R_E = \frac{V_{E1}}{I_{E1}} \approx \frac{V_{E1}}{I_C}$$

For maximum output swing, assume the voltage at the collector of Q_2, $V_{C2} = \frac{V_{CC}}{2}$

$$V_{RC} = V_{CC} - V_{C2}$$

$$R_C = \frac{V_{RC}}{I_C}$$

$$V_{C2} = V_{E1} + V_{CE1} + V_{CE2}$$

$$V_{CE1} + V_{CE2} = V_{C2} - V_{E1}$$

Assuming, $V_{CE1} = V_{CE2} = V_{CE}$, $V_{CE} = \frac{V_{C2} - V_{E1}}{2}$

Voltage at the base of Q_1, $V_{B1} = V_{BE1} + V_{E1}$

Voltage at the base of Q_2, $V_{B2} = V_{BE2} + V_{CE1} + V_{E1}$

$$V_{R3} = V_{B1}$$

$$V_{R2} = V_{B2} - V_{B1}$$

$$V_{R1} = V_{CC} - V_{B2}$$

For better stability, $R_3 << \beta R_E$, *i.e.*, $R_3 = \frac{\beta R_E}{10}$

The current through R_3, $I = \dfrac{V_{R3}}{R_3}$

Assuming same current flows through R_2 and R_1,

$$R_2 = \frac{V_{R2}}{I} \quad \text{and} \quad R_1 = \frac{V_{R1}}{I}$$

$$A_V = \frac{V_o}{V_s} = -g_{m1}g_{m2}\left[\frac{r_{\pi 2}}{1+\beta_2}\right]R_C \approx -g_{m1}R_C$$

$$A_V \approx \frac{-R_C}{r_{e1}+R_{E1}} \quad \text{and} \quad r_e = \frac{V_T}{I_E} \approx \frac{V_T}{I_C}$$

$$R_{E2} = R_E - R_{E1}$$

Input resistance $R_i = R_2 \,||\, R_3 \,||\, \beta(r_{e1}+R_{E1})$

Output resistance $R_o = r_{o2} \,||\, R_C$

$$r_{o2} \approx \frac{V_A}{I_C}$$

Design Steps

Assume $V_{E1} = 1/10^{\text{th}}$ of $V_{CC} = 1.5$ V and $I_E \approx I_C$

$$R_E = 0.5\ \text{k}\Omega, \quad V_{C2} = 7.5\ \text{V}$$

$$V_{RC} = 7.5\ \text{V}, \quad R_C = 2.5\ \text{k}\Omega$$

$$V_{CE1} = V_{CE2} = V_{CE} = 3\ \text{V}$$

$$V_{B1} = V_{R3} = 2.2\ \text{V} \quad \text{and} \quad V_{B2} = 5.2\ \text{V}$$

$$V_{R2} = 3\ \text{V} \quad \text{and} \quad V_{R1} = 9.8\ \text{V}$$

$$R_3 = 8.75\ \text{k}\Omega$$

$$I = 0.25\ \text{mA}$$

$$R_2 = 12\ \text{k}\Omega \quad \text{and} \quad R_1 = 39.2\ \text{k}\Omega$$

$$r_{e1} = r_{e2} = 8.6\ \Omega$$

$$A_V = -50 = -\frac{2500}{8.6+R_{E1}}$$

$$R_{E1} = 41.4\ \Omega \quad \text{and} \quad R_{E2} = 458.6\ \Omega$$

Input resistance $R_i = 3.21\ \text{k}\Omega$

$$r_{o2} = 66.67\ \text{k}\Omega$$

Output resistance $R_o = 2.41\ \text{k}\Omega$

Choose $C_{C1} = C_{C2} = C_E = C_B = 10\ \mu\text{F}$

3.8 DIFFERENTIAL AMPLIFIER WITH DOUBLE-ENDED OUTPUT

Design Specifications

(*i*) Differential gain $A_d = 40$ dB (*ii*) Input resistance $R_i \geq 200\ \text{k}\Omega$

(*iii*) Differential-mode input voltage $|V_{id}| \leq 40$ mV (*iv*) Input common-mode voltage range $V_{ic} = \pm 5$ V minimum.

Specify a current source to give CMRR of at least 80 dB for a single-ended output.

Use Si NPN transistors 2N2222 having $\beta_{min} = 100$, $\beta_{typ} = 175$, and $V_A = 200$.

Assume $V_{BE} = 0.7$ V.

Theory

Figure 3.8 shows the basic BJT differential-pair configuration with double-ended output. Two identical transistors, Q_1 and Q_2, whose emitters are connected together, are biased by a constant current source I_{EE}, which is connected to a negative supply voltage $-V_{EE}$. The collectors of Q_1 and Q_2 are connected through resistors R_C to a positive supply voltage V_{CC}. The transistors Q_1 and Q_2 are biased to be operated in the forward-active region. Assume that the two collector resistors R_C are equal, and that the input common-mode voltage V_{ic} sources are ideal.

Since both positive and negative bias voltages are used in the circuit, the need for coupling capacitors and voltage divider biasing resistors at the inputs of Q_1 and Q_2 has been eliminated. If the input signal voltages V_{ic} are both zero, Q_1 and Q_2 are still biased in the active region by the current source I_{EE}. Although the differential amplifier contains two transistors, it is considered a single-stage amplifier. The analysis will show that it has characteristics similar to those of the common-emitter amplifier.

When a common-mode voltage V_{ic} is applied as shown in figure to the base terminals, the transistors are biased "on" by the constant-current source, and the voltage at the common emitters is $V_E = V_{ic} - V_{BE}(on)$. Since Q_1 and Q_2 are matched or identical, current I_{EE} splits evenly between the two transistors, and

$$i_{E1} = i_{E2} = \frac{I_{EE}}{2}$$

If base currents are negligible, then $i_{C1} \approx i_{E1}$ and $i_{C2} \approx i_{E2}$ and

$$v_{C1} = V_{CC} - \left(\frac{I_{EE}}{2}\right) R_C = v_{C2}$$

This equation indicates that for an applied common-mode voltage, I_{EE} splits evenly between Q_1 and Q_2 and the difference between v_{C1} and v_{C2} is zero.

A voltage difference is created between v_{C2} and v_{C1} when a differential-mode input voltage V_{id} is applied. The differential gain, $A_d = g_m R_C$ and the input resistance, $R_i = 2\, r_\pi$.

Circuit Diagram

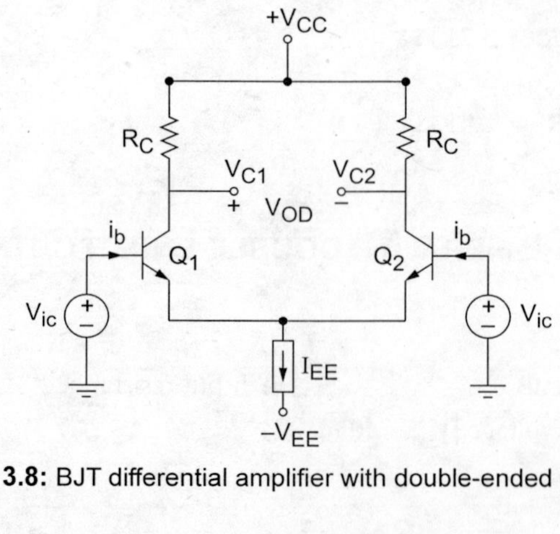

Fig. 3.8: BJT differential amplifier with double-ended output

Design Equations

$$g_m = \frac{I_C}{V_T}, \quad A_d = g_m R_C$$

The input resistance of a differential amplifier $R_i = 2r_\pi$

$$I_C \le \frac{\beta V_T}{r_\pi}$$

$V_{CC} \ge V_{IC} + v_C$, where v_C is the maximum dc + ac signal at each collector.

The resistance R_{EE} of the current source I_{EE} is given as $R_{EE} \ge \dfrac{CMRR}{g_m}$

Design Steps

$$g_m = \frac{I_C}{26 \text{ mV}}$$

$$A_d = 40 \text{ dB} = 100$$

$$A_d = \frac{I_C R_C}{26 \text{ mV}}$$

Voltage drop across $R_C = I_C R_C = 2.6$ V

The input resistance of a differential amplifier $R_i = 2r_\pi$

Taking $R_i = 200 \text{ k}\Omega, \quad r_\pi = 100 \text{ k}\Omega$

This requires $I_C \le 45.5\ \mu$A

Choose $I_C = 40\ \mu$A

$R_C = 65 \text{ k}\Omega$

As $V_{ic} = \pm 5$ V, the collector voltage is to be at least 5 V always.

Since $|V_{id}| \le 40$ mV, $V_{od} = 40 \times 100$ mV = 4 V. Therefore 4V/2 = 2V appears at each collector. The maximum dc + ac signal at each collector, $v_C = 5/2 + 2 = 4.5$ V.

$$V_{CC} \ge 5 + 4.5 \text{ V} = 9.5 \text{ V}$$

Choose $V_{CC} = 10$ V

Assume a symmetrical supply $V_{EE} = -10$ V.

$$R_{EE} \ge \frac{10000}{(40\ \mu\text{A}/26 \text{ mV})} = 6.5 \text{ M}\Omega$$

$I_{EE} = 2I_C = 80\ \mu$A with a resistance R_{EE} of 10 MΩ will provide some design margin.

3.9 DIFFERENTIAL AMPLIFIER WITH SINGLE-ENDED OUTPUT

Design Specifications

(*i*) Differential gain $A_d = 50$

(*ii*) Common-mode gain $A_c \le 1$

(*iii*) Supply voltages $V_{CC} = |V_{EE}| = 12$ V

(*iv*) Collector current $I_C = 0.5$ mA.

Use Si NPN transistors 2N2222 having $\beta_{min} = 100$, $\beta_{typ} = 175$, and $V_A = 200$.

Assume $V_{BE} = 0.7$ V. Find CMRR and R_i of the circuit.

Theory

Figure 3.9 shows the basic BJT differential-pair configuration with single-ended output. Two identical transistors, Q_1 and Q_2, whose emitters are connected together through emitter resistors R_E, are biased by a constant current source having resistance R, which is connected to a negative supply voltage $-V_{EE}$. The collector of Q_2 is connected through resistor R_C to a positive supply voltage V_{CC}. As shown, the circuit provides a single-ended output V_o at the collector of Q_2. The transistors Q_1 and Q_2 are biased to be operated in the forward-active region. The input signals at the base terminals of Q_1 and Q_2 are V_{i1} and V_{i2}, respectively.

Since both positive and negative bias voltages are used in the circuit, the need for coupling capacitors and voltage divider biasing resistors at the inputs of Q_1 and Q_2 has been eliminated. Although the differential amplifier contains two transistors, it is considered a single-stage amplifier. The analysis will show that it has characteristics similar to those of the common-emitter amplifier. The tail current provided by the circuit is

$$I_{tail} = 2I_C$$

The differential and common-mode gains are

$$A_d = \frac{R_C}{2(r_e + R_E)}$$

$$|A_c| = \frac{R_C}{r_e + R_E + 2R}$$

where r_e is the emitter resistance.

The common-mode rejection ratio $CMRR = \left|\frac{A_d}{A_c}\right| = \frac{R}{r_e + R_E}$, since $R >> (r_e + R_E)$

The input resistance of the circuit $R_i = 2\beta(r_e + R_E)$

Circuit Diagram

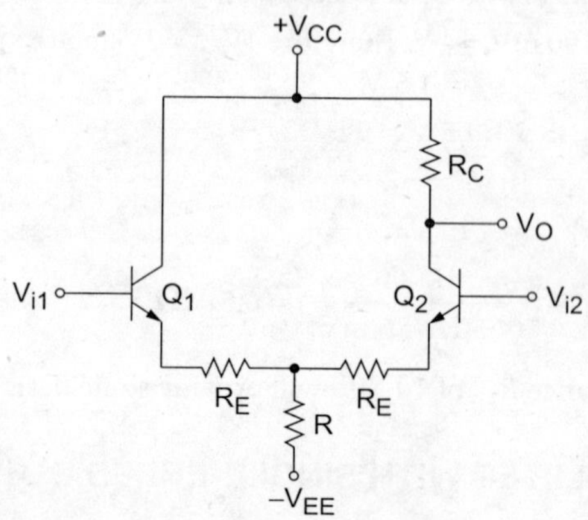

Fig. 3.9: BJT differential amplifier with single-ended output

Design Equations

$$V_o = V_{CC} - I_C R_C$$

$$\text{Tail current } I_{tail} = 2I_C$$

$$r_e = \frac{V_T}{I_C}$$

Differential gain $A_d = \dfrac{R_C}{2(r_e + R_E)}$

Common-mode gain $|A_c| = \dfrac{R_C}{r_e + R_E + 2R}$

$$CMRR = \left|\frac{A_d}{A_c}\right| = \frac{R}{r_e + R_E}$$

$$R_i = 2\beta(r_e + R_E)$$

Design Steps

Choose R_C to center V_o for maximum output swing range of 0 to 12 V, *i.e.*, $V_o = \dfrac{V_{CC}}{2} = 6$ V.

$$R_C = \frac{V_{CC} - V_o}{I_C} = 12\ \text{k}\Omega$$

Tail current $I_{tail} = 1$ mA

$$I_{tail} = \frac{0-(-12)}{R}$$

$$R = 12\ \text{k}\Omega$$

$$r_e = 51.6\ \Omega$$

Differential gain $A_d = 50 = \dfrac{12000}{2(51.6 + R_E)}$

$$R_E = 68\ \Omega$$

Common-mode gain $|A_c| = 0.4975 < 1$ required.

$$CMRR = 100.5 = 40.04\ \text{dB}$$

CMRR is also equal to $\dfrac{R}{r_e + R_E} = 100.33 = 40.03$ dB

$$R_i = 2\beta(r_e + R_E) = 41.86\ \text{k}\Omega$$

CMRR can be improved by decreasing $(r_e + R_E)$. This increases I_C.

CMRR can also be improved by increasing *R*.

3.10 SIMPLE CURRENT SOURCE

Design Specifications

(*i*) Output current = 1 mA (*ii*) Supply voltage $V_{CC} = 15$ V.

Use Si NPN transistors 2N2222 having $\beta_{min} = 100$, $\beta_{typ} = 175$, $I_S = 3.295 \times 10^{-14}$ A, and $V_A = 200$ V. Assume, $V_{BE1} = V_{BE2} = V_{CE1} = 0.7$ V.

Find the output resistance R_o, Thevenin's equivalent voltage V_{Th}, and the output current ratio if $V_{CE2} = 10$ V.

Theory

Figure shows the circuit of a simple current source using BJT. Here, the non-zero base current of the BJT causes an error in the current transfer ratio, determined by the relative areas of the emitter-base junctions of Q_1 and Q_2.

When β is sufficiently high, the base currents can be neglected. The reference current I_{REF} is passed through the diode-connected transistor Q_1 and thus establishes a corresponding voltage, V_{BE}, which in turn is applied between base and emitter of Q_2. Now, if Q_2 is matched to Q_1 or, more specifically, if the EBJ area of Q_2 is same as that of Q_1, then I_s of Q_1 and Q_2 are same. Also, the collector current of Q_2 will be equal to that of Q_1. Thus,

$$I_0 = I_R$$

For this to happen, however, Q_2 must be operating in the active mode, which in turn is achieved so long as the collector voltage V_O is 0.3 V or so higher than that of the emitter. To obtain a current transfer ratio other than unity, say k, the area of EBJ of Q_2 is made k times that of Q_1. In this case,

$$I_0 = kI_R$$

In general, the current transfer ratio is given by

$$\frac{I_O}{I_R} = \frac{I_{S2}}{I_{S1}} = \frac{\text{Area of the EBJ of } Q_2}{\text{Area of the EBJ of } Q_1}$$

Since Q_1 and Q_2 are matched and have the same V_{BE}, their collector currents will be equal. The node equation at the collector of Q_1 yields

$$I_R = I_C + \frac{2I_C}{\beta} = I_C\left(1+\frac{2}{\beta}\right)$$

Finally, since $I_O = I_{REF}$, the current transfer ratio can be found as

$$\frac{I_O}{I_R} = \frac{I_C}{I_C\left(1+\frac{2}{\beta}\right)} = \frac{1}{1+\frac{2}{\beta}}$$

As β approaches, I_O/I_{REF} approaches the nominal value of unity. For a mirror with a nominal current transfer ratio k, that is, one in which $I_{S2} = kI_{S1}$, the actual current transfer ratio is given by

$$\frac{I_O}{I_R} = \frac{k}{1+(k+1)/\beta}$$

The output resistance R_O of the current source is

$$R_O = \frac{\Delta V_O}{\Delta I_O} = r_{O2} = \frac{V_{A2}}{I_O}$$

where V_{A2} and r_{o2} are the Early voltage and the output resistance, respectively, of Q_2.

Circuit Diagram

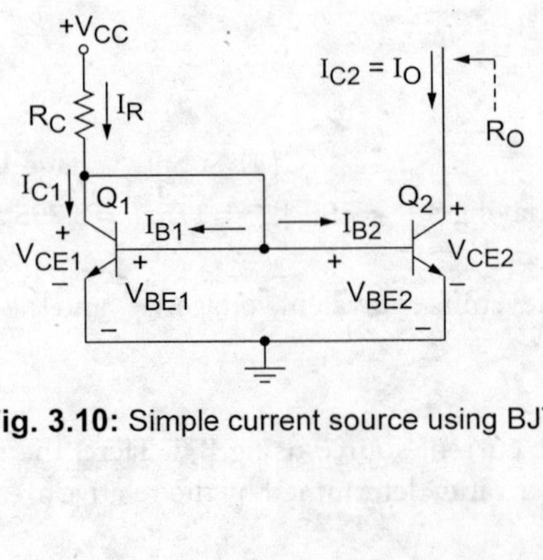

Fig. 3.10: Simple current source using BJT

Design Equations

$$I_O = I_{C2} = I_{C1} = \frac{I_R}{1+2/\beta}$$

$$R = \frac{V_{CC} - V_{BE1}}{I_R}$$

$$R_O = \frac{V_A}{I_{C2}} = \frac{V_A}{I_O}$$

$$V_{Th} = I_{C2}R_O = V_A$$

$$\frac{I_{C2}}{I_{C1}} = \frac{1+V_{CE2}/V_A}{1+V_{CE1}/V_A}$$

Design Steps

$$I_O = I_{C2} = I_{C1} = 1 \text{ mA}$$

$$I_R = I_O\left(1+\frac{2}{\beta}\right) = 1.011 \text{ mA}$$

$$R = 14.14 \text{ k}\Omega$$

$$R_O = 200 \text{ k}\Omega$$

$$V_{Th} = 200 \text{ V}$$

$$V_{CE1} = V_{CC} - I_R R = 0.7045 \text{ V}$$

For $V_{CE2} = 10$ V,

$$\frac{I_{C2}}{I_{C1}} = 1.046$$

3.11 VOLTAGE-SERIES (SHUNT-SERIES) FEEDBACK AMPLIFIER

Design Specifications

(*i*) Closed – loop voltage gain $\frac{V_o}{V_S} = 0.9$

(*ii*) The Q-point is to be located at $I_C = 3$ mA and $V_{CE} = 6$ V

(*iii*) Signal source resistance $R_S = 250\ \Omega$.

Use Si NPN transistor 2N2222 having $\beta_{min} = 100$, $\beta_{typ} = 175$, and $V_A = 200$ V.

Theory

The emitter-follower circuit shown in Fig. 3.11 provides voltage-series feedback. The signal voltage V_s is the input voltage. The output voltage V_o is also the feedback voltage in series with the input voltage. C_{C1} and C_{C2} are the input and output coupling capacitors, respectively. Let V_i is the input voltage at the base of transistor and V_f is the voltage across R_E and is also the feedback voltage. Considering R_S as part of the amplifier $V_i = V_s$ and the voltage gain without feedback is

$$A_V = \frac{V_o}{V_i} = \frac{\beta I_b R_E}{V_s} = \frac{\beta R_E}{R_S + r_\pi}$$

The feedback factor $\beta' = \dfrac{V_o}{V_f} = 1$

The gain with feedback $A_{Vf} = \dfrac{V_o}{V_s} \approx \dfrac{\beta R_E}{R_S + r_\pi + \beta R_E}$

Circuit Diagram

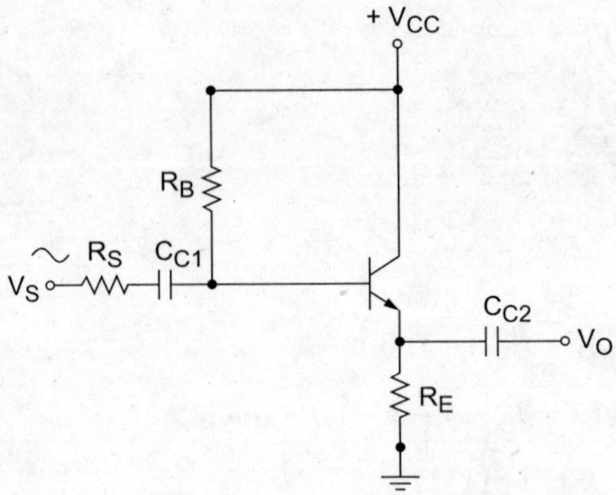

Fig. 3.11: BJT voltage-series (shunt-series) feedback amplifier

Design Equations

$$V_{CC} \approx 2V_{CE}$$

$$I_B = \frac{I_C}{\beta} \quad \text{and} \quad \alpha = \frac{\beta}{1+\beta}$$

$$I_E = \frac{I_C}{\alpha}$$

$$g_m = \frac{I_C}{V_T}$$

$$r_\pi = \frac{\beta}{g_m}$$

$$r_o \approx \frac{V_A}{I_C}$$

$$A_{Vf} \approx \frac{\beta R_E}{R_S + r_\pi + \beta R_E}$$

The voltage at the base, $V_B = V_{BE} + V_E$

$$R_B = \frac{V_{CC} - V_B}{I_B}$$

Design Steps

The circuit is an emitter follower.

Choose, $V_{CC} = 2V_{CE} = 12$ V

$I_B = 0.017$ mA

$\alpha = 0.994$

$I_E = 3.017$ mA

$g_m = 116.3$ mA/V

$r_\pi = 1.5$ kΩ

$r_o = 66.67$ kΩ

Generally, R_B is very large as compared to r_π. Also, r_o is large. Neglecting R_B and r_o,

$$A_{Vf} = 0.9 \approx \frac{\beta R_E}{R_S + r_\pi + \beta R_E}$$

$R_E = 90\ \Omega$

The voltage at the base $V_B = 0.9715$ V

The value of $R_B = 648.74$ kΩ

Choose $C_{C1} = C_{C2} = 10\ \mu$F, 16 V

3.12 VOLTAGE-SERIES (SHUNT-SERIES) FEEDBACK PAIR

Design Specifications

(*i*) Closed – loop voltage gain, $\frac{V_o}{V_S} = 50$

(*ii*) The Q-point is to be located at $I_C = 3$ mA and $V_{CE} = 6$ V

(*iii*) Neglect the signal source resistance.

Use Si NPN transistors 2N2222 having $\beta_{min} = 100$, $\beta_{typ} = 175$, and $V_A = 200$ V.

Theory

The circuit in Fig. 3.12 shows a two-stage amplifier which makes use of voltage-series feedback by connecting the second collector to the first emitter through the voltage divider R_3-R_4. Capacitors C_C are dc blocking capacitors, and capacitors C_E are bypass capacitors for the emitter bias resistors. All these capacitances represent negligible reactance at the frequencies of operation of this circuit. For this amplifier, the voltage gain A_{Vf} is given approximately by $1/\beta'$ and is thus stabilized against temperature changes and transistor replacement.

The closed-loop gain or gain with feedback,

$$A_{Vf} = \frac{A_V}{1+\beta' A_V}$$

where β' = feedback factor = $\frac{R_4}{R_3 + R_4}$ and A_V is the open-loop voltage gain or gain without feedback and is given by

$$A_V = \left[\frac{-\beta(r_o \,||\, R_C \,||\, R_1 \,||\, R_2 \,||\, r_\pi)}{r_\pi + (1+\beta)R_E}\right]\cdot\left[\frac{-\beta\{r_o \,||\, R_C \,||\, (R_3 + R_4)\}}{r_\pi}\right]$$

Circuit Diagram

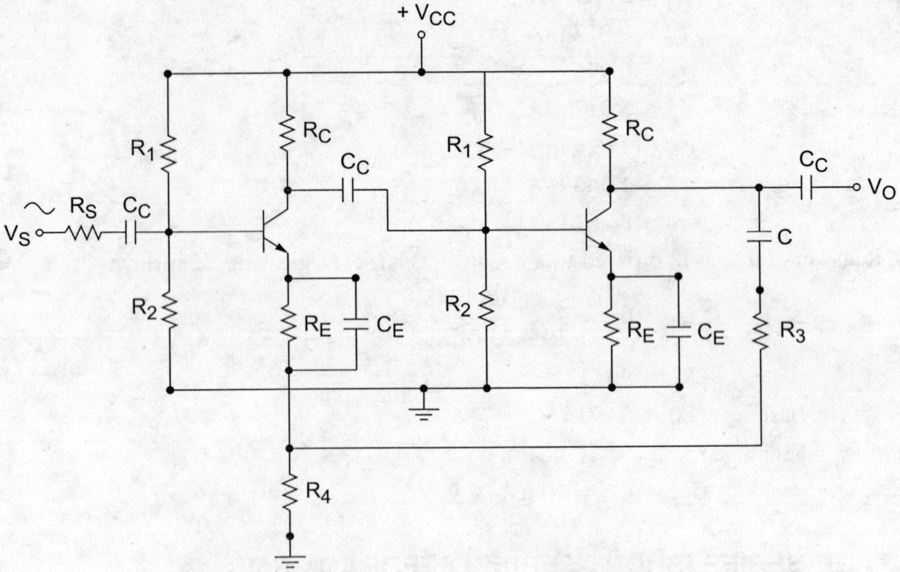

Fig. 3.12: BJT voltage-series (shunt-series) feedback pair

Design Equations

$$V_{CC} \approx 2V_{CE}$$

$$I_B = \frac{I_C}{\beta}$$

$$\alpha = \frac{\beta}{1+\beta}$$

$$I_E = \frac{I_C}{\alpha}$$

$$V_E \approx 1/10^{\text{th}} \text{ of } V_{CC}$$

Voltage across R_C, $V_{RC} = V_{CC} - V_{CE} - V_E$

Equivalent resistance at the base of transistor, $R_B = \dfrac{(1+\beta)R_E}{10}$

Thevenin voltage at the base terminal, $V_{Th} = R_B I_B + V_{BE} + V_E$

$$R_1 = \frac{R_B V_{CC}}{V_{Th}}$$

$$R_2 = \frac{R_B V_{CC}}{V_{CC} - V_{Th}}$$

$$g_m = \frac{I_C}{V_T}$$

$$r_\pi = \frac{\beta}{g_m}$$

$$r_o \approx \frac{V_A}{I_C}$$

The closed-loop gain, $A_{Vf} = \dfrac{A_V}{1+\beta' A_V}$, β' = feedback factor $= \dfrac{R_4}{R_3 + R_4}$

$$A_V = \left[\frac{-\beta(r_o \,||\, R_C \,||\, R_1 \,||\, R_2 \,||\, r_\pi)}{r_\pi + (1+\beta)R_E}\right] \cdot \left[\frac{-\beta\{r_o \,||\, R_C \,||\, (R_3 + R_4)\}}{r_\pi}\right]$$

Since R_4 is part of R_E of 1st stage, choose $R_4 << R_E$, *i.e.*, $R_4 = \dfrac{R_E}{10}$

Design Steps

Design of biasing circuit:

Choose $V_{CC} = 2V_{CE} = 12$ V

$I_B = 0.017$ mA

$\alpha = 0.994$

$I_E = 3.017$ mA

Taking emitter voltage as 1/10th of V_{CC}, $V_E = 1.2$ V

$R_E = 398\ \Omega$

Voltage across R_C, $V_{RC} = 4.8$ V

$R_C = 1.6$ kΩ

Equivalent resistance at the base of transistor, $R_B = 7$ kΩ

Thevenin voltage at the base terminal, $V_{Th} = 2.02$ V

For the circuit to operate efficiently, it is assumed that the current through R_1 and R_2 is approximately equal and much larger than the base current.

$R_1 = 41.58$ kΩ

$R_2 = 8.42$ kΩ

$g_m = 116.3$ mA/V

$r_\pi = 1.5$ kΩ

$r_o = 66.67$ kΩ

Design of Feedback Circuit

Choose $R_4 = \dfrac{R_E}{10} \approx 40\ \Omega$

Substituting the known values in A_V, A_{Vf} and simplifying,

$R_3 = 2.85$ kΩ

To get the required gain of 50 exactly, R_3 can be taken as a series combination of 2 kΩ and 1 kΩ potentiometer.

Choose, $C_C = C = 10\ \mu$F, 16 V and $C_E = 50\ \mu$F, 16 V

3.13 CURRENT-SERIES (SERIES-SERIES) FEEDBACK AMPLIFIER

Design Specifications

(*i*) Closed – loop voltage gain $\frac{V_o}{V_S} = 8$ (*ii*) Collector current $I_C = 3$ mA

(*iii*) Signal source resistance $R_S = 250\ \Omega$.

Use Si NPN transistors 2N2222 having $\beta_{min} = 100$, $\beta_{typ} = 175$, and $V_A = 200$ V.

Theory

The circuit shown in Fig. 3.13 is a *CE* transistor amplifier with emitter resistor R_E. The feedback signal is the voltage V_f across R_E and the sampled signal is the load current I_O. Hence this is a case of current-series feedback. The base current compared with collector or load current can be neglected.

Although I_O is proportional to V_o it is not correct to conclude that this is a voltage-series feedback. Thus, if the output signal is taken as the voltage V_o then the feedback factor,

$$\beta' = \frac{V_f}{V_o} = -\frac{I_O R_E}{I_O R_C} = -\frac{R_E}{R_C}$$

Since β' is now a function of the load R_C, the sampled signal is not output voltage and the circuit topology is not voltage-series type.

The voltage gain with feedback, $A_{Vf} = \dfrac{-\beta(r_o \,||\, R_C)}{R_S + \{R_B \,||\, (r_\pi + (1+\beta)R_E)\}}$

Circuit Diagram

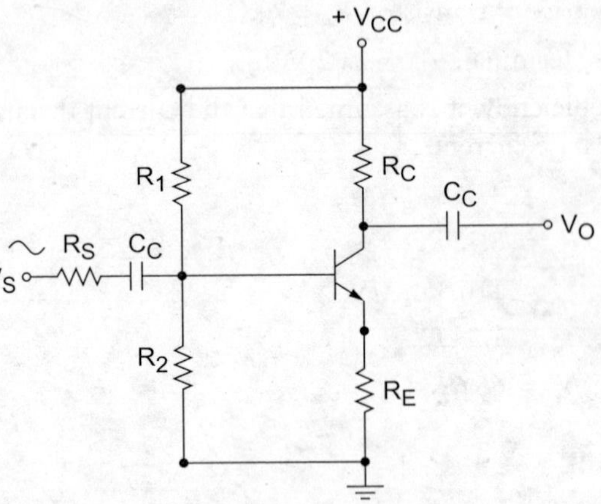

Fig. 3.13: BJT current-series (series-series) feedback amplifier

Design Equations

$$I_B = \frac{I_C}{\beta} \quad \text{and} \quad \alpha = \frac{\beta}{1+\beta}$$

$$I_E = \frac{I_C}{\alpha}$$

$$V_E \approx 1/10^{th} \text{ of } V_{CC}$$

Equivalent resistance at the base of transistor, $R_B = \dfrac{(1+\beta)R_E}{10}$

Thevenin voltage at the base terminal, $V_{Th} = R_B I_B + V_{BE} + V_E$

$$R_1 = \frac{R_B V_{CC}}{V_{Th}} \quad \text{and} \quad R_2 = \frac{R_B V_{CC}}{V_{CC} - V_{Th}}$$

$$g_m = \frac{I_C}{V_T}$$

$$r_\pi = \frac{\beta}{g_m} \quad \text{and} \quad r_o \approx \frac{V_A}{I_C}$$

The closed-loop gain $A_{Vf} = \dfrac{-\beta(r_o \,||\, R_C)}{R_S + \{R_B \,||\, (r_\pi + (1+\beta)R_e)\}}$

$$V_{CE} = V_{CC} - I_C R_C - V_E$$

Design Steps

Design of Biasing Circuit

Choose

$$V_{CC} = 12 \text{ V}$$
$$I_B = 0.017 \text{ mA}$$
$$\alpha = 0.994$$
$$I_E = 3.017 \text{ mA}$$

Taking emitter voltage as $1/10^{th}$ of V_{CC}, $V_E = 1.2$ V

$$R_E = 398\ \Omega$$

Equivalent resistance at the base of transistor, $R_B = = 7$ kΩ

Thevenin voltage at the base terminal, $V_{Th} = 2.02$ V

For the circuit to operate efficiently, it is assumed that the current through R_1 and R_2 is approximately equal and much larger than the base current.

$$R_1 = 41.58 \text{ k}\Omega$$
$$R_2 = 8.42 \text{ k}\Omega$$
$$g_m = 116.3 \text{ mA/V}$$
$$r_\pi = 1.5 \text{ k}\Omega$$
$$r_o = 66.67 \text{ k}\Omega$$

Design of Feedback Circuit

Substituting the known values in A_{Vf} expression and simplifying,

$$R_C = 304\ \Omega$$
$$V_{CE} = 9.888 \text{ V}$$

To get the required gain of 8 exactly, R_E can be taken as a series combination of 39 Ω and 1 kΩ potentiometer.

Choose $C_C = 10$ µF, 16 V

3.14 VOLTAGE-SHUNT (SHUNT-SHUNT) FEEDBACK AMPLIFIER

Design Specifications

(*i*) Closed – loop voltage gain $\frac{V_o}{V_S} = 8$

(*ii*) The Q-point is to be located at $I_C = 3$ mA and $V_{CE} = 6$V

(*iii*) Signal source resistance $R_S = 250\ \Omega$.

Use Si NPN transistors 2N2222 having $\beta_{min} = 100$, $\beta_{typ} = 175$, and $V_A = 200$ V.

Theory

Figure 3.14 shows a common-emitter stage with a resistor R_F connected from the output to the input. In the circuit shown, the output voltage V_o is much greater than the input voltage V_{in} and is 180° out of phase with V_{in}. The feedback current through R_F is

$$I_f = \frac{V_s - V_o}{R_F}$$

The feedback factor $\beta' = -1/R_F$. Since the feedback current is proportional to the output voltage, this circuit is an example of a voltage-shunt feedback amplifier.

The transresistance $R_{Mf} = \frac{V_o}{I_S} = \frac{-1}{\beta'} = R_F$

The voltage gain with feedback, $|A_{VF}| \approx \frac{R_F}{R_S}$

Circuit Diagram

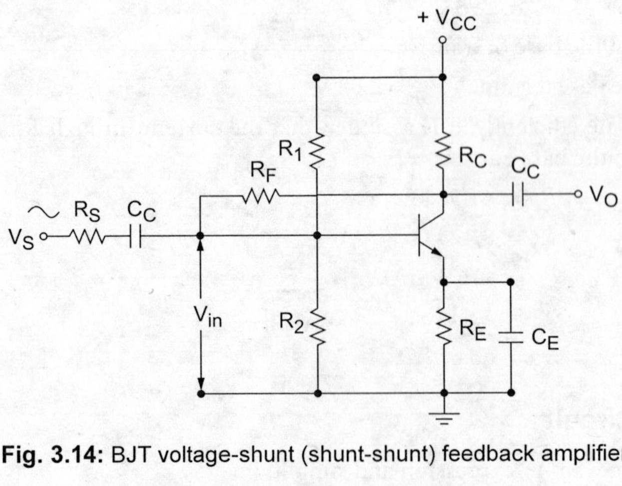

Fig. 3.14: BJT voltage-shunt (shunt-shunt) feedback amplifier

Design Equations

$$V_{CC} \approx 2V_{CE}$$

$$I_B = \frac{I_C}{\beta} \quad \text{and} \quad \alpha = \frac{\beta}{1+\beta}$$

$$I_E = \frac{I_C}{\alpha}$$

$$V_E \approx 1/10^{th} \text{ of } V_{CC}$$

Voltage across R_C, $\quad V_{RC} = V_{CC} - V_{CE} - V_E$

Equivalent resistance at the base of transistor, $R_B = \dfrac{(1+\beta)R_E}{10}$

Thevenin voltage at the base terminal, $V_{Th} = R_B I_B + V_{BE} + V_E$

$$R_1 = \frac{R_B V_{CC}}{V_{Th}} \quad \text{and} \quad R_2 = \frac{R_B V_{CC}}{V_{CC} - V_{Th}}$$

The voltage gain with feedback, $|A_{VF}| \approx \dfrac{R_F}{R_S}$

Design Steps

Design of Biasing Circuit

Choose

$$V_{CC} = 2V_{CE} = 12 \text{ V}$$

$$I_B = 0.017 \text{ mA}$$

$$\alpha = 0.994$$

$$I_E = 3.017 \text{ mA}$$

Taking emitter voltage as $1/10^{th}$ of V_{CC}, $V_E = 1.2$ V

$$R_E = 398\ \Omega$$

Voltage across R_C, $\quad V_{RC} = 4.8$ V

$$R_C = 1.6\ \text{k}\Omega$$

Equivalent resistance at the base of transistor, $R_B = 7\ \text{k}\Omega$

Thevenin voltage at the base terminal, $V_{Th} = 2.02$ V

For the circuit to operate efficiently, it is assumed that the current through R_1 and R_2 is approximately equal and much larger than the base current.

$$R_1 = 41.58\ \text{k}\Omega$$

$$R_2 = 8.42\ \text{k}\Omega$$

Feedback Circuit Design

The voltage gain with feedback, $|A_{VF}| \approx 8$

$$R_F = 8R_S = 2\ \text{k}\Omega$$

To get the required gain of 8 exactly, R_F can be taken as a series combination of 1.5 kΩ and 1 kΩ potentiometer.

Choose, $\quad C_C = 10\ \mu\text{F}, 16\text{ V}$ and $C_E = 50\ \mu\text{F}, 16\text{ V}$

3.15 CURRENT-SHUNT (SERIES-SHUNT) FEEDBACK AMPLIFIER

Design Specifications

(*i*) Closed – loop voltage gain $\dfrac{V_o}{V_S} = 8$

(*ii*) The Q-point is to be located at $I_C = 3$ mA and $V_{CE} = 6$ V

(*iii*) Signal source resistance $R_S = 250\ \Omega$.

Use Si NPN transistors 2N2222 having $\beta_{min} = 100$, $\beta_{typ} = 175$, and $V_A = 200$ V.

Theory

Figure 3.15 shows two transistors in cascade, with feedback from the second emitter to the first base through the resistor R_F. The voltage V_{i2} at the base of Q_2 is much larger than V_{i1} at the base of Q_1 because of the voltage gain of Q_1. Also, V_{i2} is 180° out of phase with V_{i1}, Because of emitter-follower action, the voltage V_{E2} across R_{E2} is only slightly smaller than V_{i2}, and these voltages are in phase. Hence V_{E2} is larger in magnitude than V_{i1} and is 180° out of phase with V_{i1}. If the input signal increases so that the current I_S through R_S increases, the feedback current I_f flowing through R_F from input to output also increases, and the current entering into base of Q_1 is $I_i = I_S - I_f$ is smaller than it would be if there were no feedback. This action is characteristic of negative feedback. Since $V_{E2} >> V_{i1}$, and neglecting the base current of Q_2 compared with the collector current,

$$I_f = \frac{V_{i1} - V_{E2}}{R_F} \approx -\frac{V_{E2}}{R_F} = \frac{(I_O - I_f)R_{E2}}{R_F}$$

or

$$I_f = \frac{I_O R_{E2}}{R_F + R_{E2}} = \beta' I_O$$

where I_O is the current leaving the collector of Q_2 and $\beta' = \dfrac{R_{E2}}{R_F + R_{E2}}$ is the feedback factor. Since the feedback current is proportional to the output current, this circuit is an example of a current-shunt feedback amplifier. The current gain with feedback is

$$A_{If} = \frac{1}{\beta'} = \frac{R_F + R_{E2}}{R_{E2}}$$

and the voltage gain, $|A_{Vf}| = \left[\dfrac{R_F + R_{E2}}{R_{E2}}\right]\left[\dfrac{r_o \,||\, R_{C2}}{R_S}\right]$

Circuit Diagram

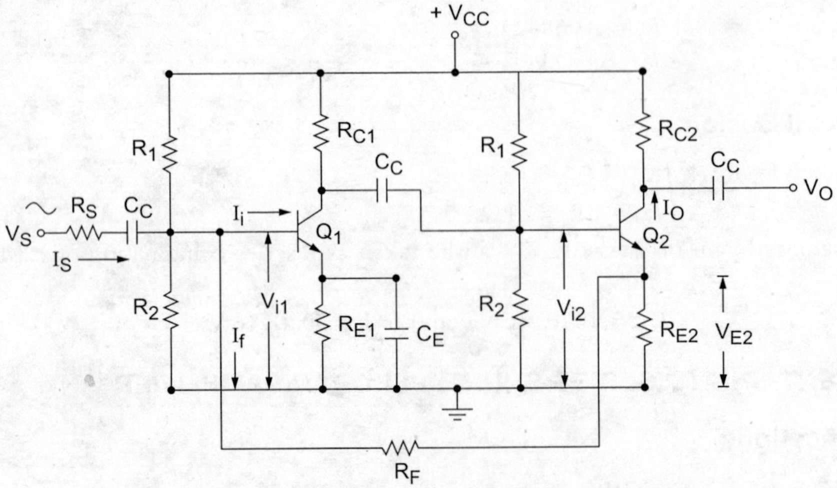

Fig. 3.15: BJT current-shunt (series-shunt) feedback amplifier

Design Equations

$$V_{CC} \approx 2V_{CE}$$

$$I_B = \frac{I_C}{\beta}$$

$$\alpha = \frac{\beta}{1+\beta}$$

$$I_E = \frac{I_C}{\alpha}$$

$$V_E \approx 1/10^{\text{th}} \text{ of } V_{CC}$$

$$R_{E1} = R_{E2} = \frac{V_E}{I_E}$$

Voltage across R_C,
$$V_{RC} = V_{CC} - V_{CE} - V_E$$

$$R_{C1} = R_{C2} = \frac{V_{RC}}{I_C}$$

Equivalent resistance at the base of transistor, $R_B = \dfrac{(1+\beta)R_E}{10}$

Thevenin voltage at the base terminal, $V_{Th} = R_B I_B + V_{BE} + V_E$

$$R_1 = \frac{R_B V_{CC}}{V_{Th}}$$

$$R_2 = \frac{R_B V_{CC}}{V_{CC} - V_{Th}}$$

$$g_m = \frac{I_C}{V_T}$$

$$r_\pi = \frac{\beta}{g_m}$$

$$r_o \approx \frac{V_A}{I_C}$$

The voltage gain with feedback, $|A_{VF}| \approx \left[\dfrac{R_F + R_{E2}}{R_{E2}}\right]\left[\dfrac{r_o \,||\, R_{C2}}{R_S}\right]$

Design Steps

Design of Biasing Circuit

Choose,
$$V_{CC} = 2V_{CE} = 12 \text{ V}$$
$$I_B = 0.017 \text{ mA}$$
$$\alpha = 0.994$$
$$I_E = 3.017 \text{ mA}$$

Taking emitter voltage as $1/10^{\text{th}}$ of V_{CC}, $V_E = 1.2$ V

$$R_{E1} = R_{E2} = 398\ \Omega$$

Voltage across R_C, $V_{RC} = 4.8$ V

$$R_{C1} = R_{C2} = 1.6\ \text{k}\Omega$$

Equivalent resistance at the base of transistor, $R_B = 7\ \text{k}\Omega$

Thevenin voltage at the base terminal, $V_{Th} = 2.02$ V

For the circuit to operate efficiently, it is assumed that the current through R_1 and R_2 is approximately equal and much larger than the base current.

$$R_1 = 41.58\ \text{k}\Omega$$
$$R_2 = 8.42\ \text{k}\Omega$$
$$g_m = 116.3\ \text{mA/V}$$
$$r_\pi = 1.5\ \text{k}\Omega$$
$$r_o = 66.67\ \text{k}\Omega$$

Feedback Circuit Design

Given, the voltage gain with feedback, $|A_{VF}| = 8$

Substituting the component values in the expression for A_{VF},

$$R_F = 111\ \Omega$$

To get the required gain of 8 exactly, R_{E2} can be taken as a series combination of 100 Ω and 1 kΩ potentiometer.

Choose, $C_C = 10$ μF, 16 V and $C_E = 50$ μF, 16 V

3.16 RC PHASE-SHIFT OSCILLATOR

Design Specification

Frequency of oscillations $f_o = 2$ kHz

Theory

RC phase-shift oscillator using transistor is shown in Fig. 3.16. The output resistance of the feedback network would be shunted by the relatively low input resistance of the transistor. Hence, voltage-shunt feedback is employed in the circuit. The resistor $R' = R - R_i$, where $R_i = R_B \,||\, r_\pi$ is the input resistance of the transistor. This choice makes the three RC sections of the phase-shifting network alike. We assume that the biasing resistors R_1, R_2 and R_E have no effect on the signal operation and neglect these in the analysis. Neglecting the loading effect of feedback network consisting of R-C sections, the gain of the amplifier

$$|A| = g_m(r_o \,||\, R_C)$$

The frequency of oscillations

$$f_o = \frac{1}{2\pi RC\sqrt{6+4RC/R}}$$

For oscillations to occur, the minimum value of β of the transistor

$$\beta_{\min} = \frac{4R_C}{R} + 23 + \frac{29R}{R_C}$$

Circuit Diagram

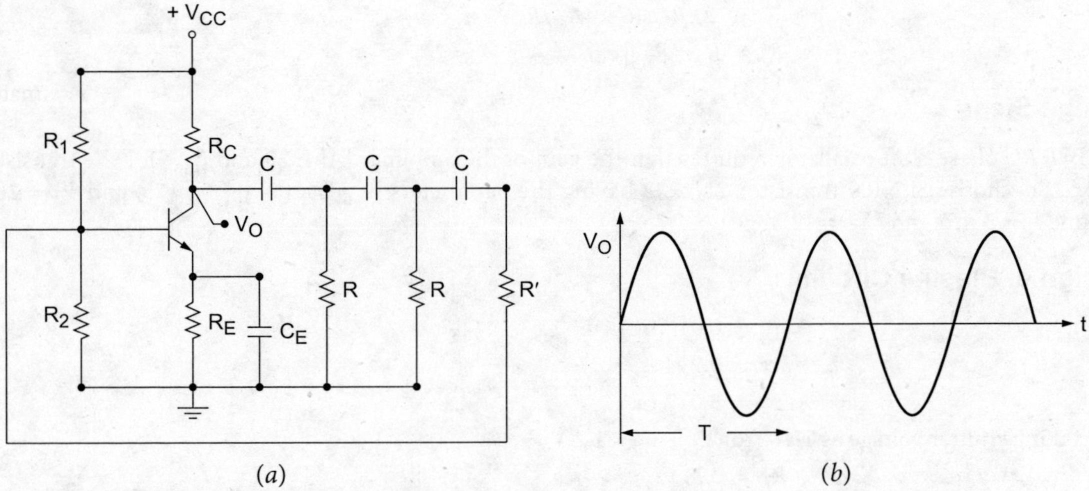

Fig. 3.16: (a) RC phase-shift oscillator using BJT and (b) output voltage waveform

Design Equations

$$I_B = \frac{I_C}{\beta}$$

$$\alpha = \frac{\beta}{1+\beta}$$

$$I_E = \frac{I_C}{\alpha}$$

$$V_E \approx 1/10^{\text{th}} \text{ of } V_{CC}$$

Equivalent resistance at the base of transistor, $R_B = \dfrac{(1+\beta)R_E}{10}$

Thevenin voltage at the base terminal, $V_{Th} = R_B I_B + V_{BE} + V_E$

$$R_1 = \frac{R_B V_{CC}}{V_{Th}}$$

$$R_2 = \frac{R_B V_{CC}}{V_{CC} - V_{Th}}$$

Neglecting the loading effect of feedback network consisting of *R-C* sections,

$$|A| \approx g_m(r_o \,||\, R_C)$$

$$g_m = \frac{I_C}{V_T}$$

$$r_o \approx \frac{V_A}{I_C}$$

$$r_\pi = \frac{\beta}{g_m}$$

The frequency of oscillations is given by,

$$f_o = \frac{1}{2\pi RC\sqrt{6+4R_C/R}}$$

$$R' = R - (R_B \,||\, r_\pi)$$

Design Steps

The BJT *RC* phase-shift oscillator requires that the gain of the amplifier $|A| \geq 29$ and β_{min} for the transistor = 44.5. So, choose Si NPN transistor 2N2222 having the parameters β_{min} = 100, β_{typ} = 175, and V_A = 200. Assume I_C = 3 mA and V_{CC} = 12 V.

Design of Biasing Circuit

$$I_B = 0.017 \text{ mA}$$

$$\alpha = 0.994$$

$$I_E = 3.017 \text{ mA}$$

Taking emitter voltage as 1/10th of V_{CC}, V_E = 1.2 V

$$R_E = 398\ \Omega$$

Equivalent resistance at the base of transistor, R_B = 7 kΩ

Thevenin voltage at the base terminal, V_{Th} = 2.02 V

For the circuit to operate efficiently, assume that the current through R_1 and R_2 is approximately equal and much larger than the base current.

$$R_1 = 41.58 \text{ k}\Omega$$

$$R_2 = 8.42 \text{ k}\Omega$$

AC Design

$$g_m = 116.3 \text{ mA/V}$$

$$r_o = 66.67 \text{ k}\Omega$$

$$r_\pi = 1.5 \text{ k}\Omega$$

Assuming the gain of the amplifier $A = -100$,

$$R_C = 871\ \Omega$$

Feedback Network Design

Choosing $C = 0.01\ \mu\text{F}, \quad R = 2.97 \text{ k}\Omega$

$$R' = R - (R_B \,||\, r_\pi) = 1.735 \text{ k}\Omega$$

In order to obtain pure sinewave, the value of R' is adjusted using a 5 kΩ potentiometer in place of fixed 1.735 kΩ resistor.

Choose $C_E = 50\ \mu\text{F, 16 V}$

3.17 WIEN BRIDGE OSCILLATOR

Design Specifications

(*i*) Frequency of oscillations f_o = 5 kHz

(*ii*) The Q-point is to be located at I_C = 3 mA and V_{CE} = 6 V.

Use Si NPN transistors 2N2222 having β_{min} = 100 and β_{typ} = 175.

Theory

The Wien bridge oscillator shown in Fig. 3.17 is essentially a two-stage amplifier with an *R-C* bridge circuit. The *R-C* bridge circuit consisting of R_1 in series with C_1, R_3, R_4, and R_2 in parallel with C_2 is a lead-lag network. The phase shift across the network lags with increasing frequency and leads with decreasing frequency. The output of transistor Q_2 is fed back to input for providing regeneration required for producing oscillations.

The bridge circuit can be used as feedback network for an oscillator, provided that the phase shift through the amplifier is zero. This requisite condition is achieved by using a two-stage amplifier, as shown in the figure. The voltage across the parallel combination R_2-C_2 is fed to the input of the first stage. The circuit uses positive and negative feedbacks to transistor Q_1. The positive feedback is provided by R_1-C_1 and R_2-C_2, while the resistors R_3 and R_4 provides negative feedback to stabilize the amplitude of the output.

The amplifier voltage gain $A = \dfrac{R_3 + R_4}{R_4} = \dfrac{R_3}{R_4} + 1 = 3$, since $R_3 = 2R_4$.

This corresponds with the feedback network attenuation of 1/3. To stabilize the gain at a very low value of 3, negative feedback is essential. Assuming $R_1 = R_2 = R$ and $C_1 = C_2 = C$, the frequency of oscillations is

$$f_o = \frac{1}{2\pi RC}$$

The continuous variation in f_o can be achieved by varying the two capacitors C_1 and C_2 simultaneously. These capacitors can be chosen as variable air-gang type. The frequency range of the oscillator can be changed by switching into the circuit different values of resistors R_1 and R_2.

Circuit Diagram

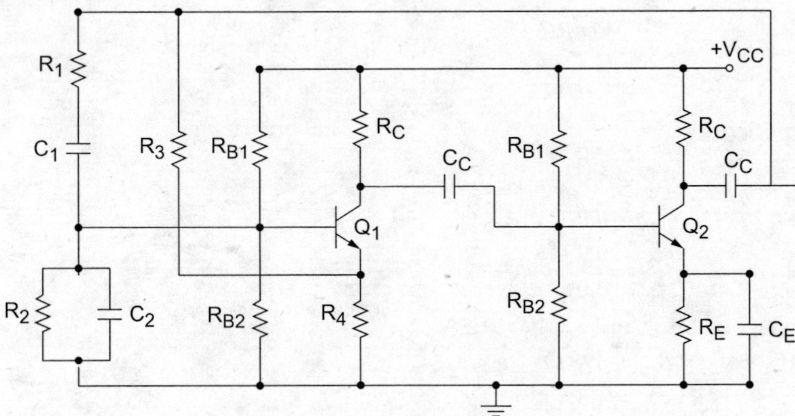

(*a*)

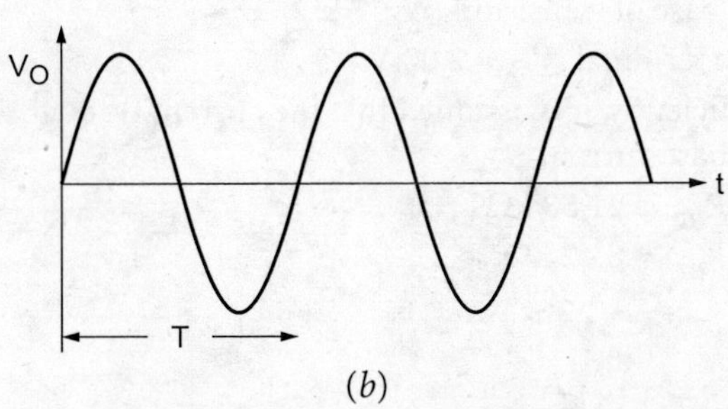

(*b*)

Fig. 3.17: (a) Wien bridge oscillator using BJT and (b) output voltage waveform

Design Equations

$$V_{CC} \approx 2V_{CE}$$

$$I_B = \frac{I_C}{\beta}$$

$$\alpha = \frac{\beta}{1+\beta}$$

$$I_E = \frac{I_C}{\alpha}$$

$$V_E \approx 1/10^{\text{th}} \text{ of } V_{CC}$$

Voltage across R_C, $V_{RC} = V_{CC} - V_{CE} - V_E$

Equivalent resistance at the base of transistor, $R_B = \dfrac{(1+\beta)R_E}{10}$

Thevenin voltage at the base terminal, $V_{Th} = R_B I_B + V_{BE} + V_E$

$$R_1 = \frac{R_B V_{CC}}{V_{Th}}$$

$$R_2 = \frac{R_B V_{CC}}{V_{CC} - V_{Th}}$$

Assuming $R_1 = R_2 = R$ and $C_1 = C_2 = C$, the frequency of oscillations is given by,

$$f_o = \frac{1}{2\pi RC}$$

Design Steps

Design of Biasing Circuit

Choose, $V_{CC} = 2V_{CE} = 12$ V

$$I_B = 0.017 \text{ mA}$$

$$\alpha = 0.994$$

$$I_E = 3.017 \text{ mA}$$

Taking emitter voltage as $1/10^{\text{th}}$ of V_{CC}, $V_E = 1.2$ V

$$R_E = 398\ \Omega$$

Voltage across R_C, $V_{RC} = 4.8$ V

$$R_C = 1.6\ \text{k}\Omega$$

Equivalent resistance at the base of transistor, $R_B = 7$ kΩ

Thevenin voltage at the base terminal, $V_{Th} = 2.02$ V

For the circuit to operate efficiently, it is assumed that the current through R_{B1} and R_{B2} is approximately equal and much larger than the base current.

$$R_{B1} = 41.58\ \text{k}\Omega$$

$$R_{B2} = 8.42\ \text{k}\Omega$$

Feedback Network Design

Taking $C = 0.01$ μF, the value of $R = 3.18$ kΩ

The condition for oscillations requires $R_3 = 2R_4$

Assuming $R_4 = 10$ kΩ, $R_3 = 20$ kΩ

In order to obtain a pure sine wave the value of R_3 is adjusted using a 10 kΩ potentiometer in series with a fixed resistor of 12 kΩ.

Choose C_C = 10 μF, 16 V and C_E = 50 μF, 16 V.

3.18 HARTLEY OSCILLATOR

Design Specifications

(*i*) Frequency of oscillations f_O = 100 kHz

(*ii*) The *Q*-point is to be located at I_C = 3 mA and V_{CE} = 6 V.

Use Si NPN transistors 2N2222 having β_{min} = 100 and β_{typ} = 175.

Theory

In the Hartley oscillator circuit shown in Fig. 3.18, the oscillation frequency is determined by a tank circuit comprising of two inductors and one capacitor. The inductors are connected in series and the capacitor is connected across them in parallel. The amplifier in *CE* configuration uses voltage-divider bias. C_C are the coupling capacitors and C_E is the emitter bypass capacitor. The phase-shift network consists of tank circuit formed by inductors L_1, L_2 and capacitor *C*.

The oscillations produced in the tank circuit are fed back to the base of transistor and appears in the amplified form across the collector and emitter of the transistor. The output voltage of the transistor will be in phase with the voltage across inductor L_1. Since the junction of two inductors is grounded, the voltage across L_2 that is actually fed back to the base will be 180° out of phase to that of the voltage across L_1, making the total phase shift in the circuit as 360°.

The frequency of oscillations is given by

$$f_o = \frac{1}{2\pi\sqrt{(L_1 + L_2)C}}$$

If mutual inductance *M* exists between L_1 and L_2, then

$$f_o = \frac{1}{2\pi\sqrt{(L_1 + L_2 + 2M)C}}$$

Circuit Diagram

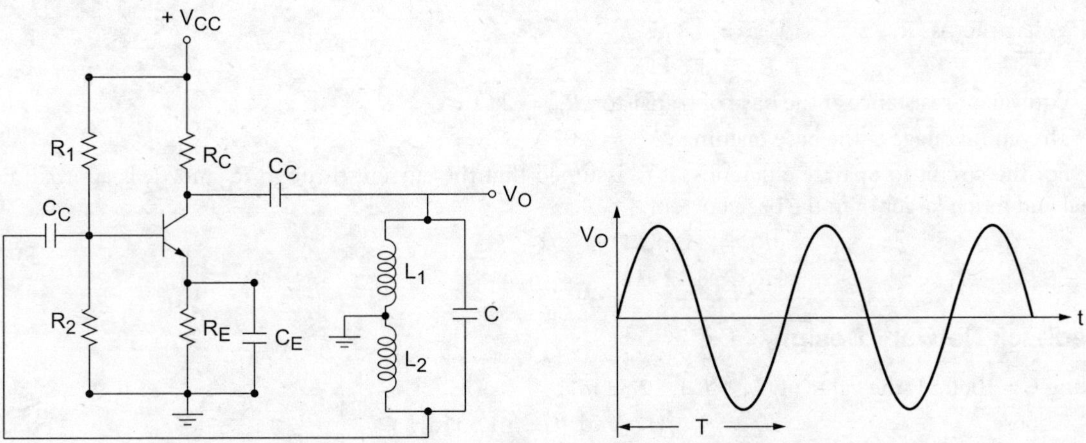

(*a*) (*b*)

Fig. 3.18: (a) Hartley oscillator using BJT and (b) output voltage waveform

Design Equations

$$V_{CC} \approx 2V_{CE}$$

$$I_B = \frac{I_C}{\beta}$$

$$\alpha = \frac{\beta}{1+\beta}$$

$$I_E = \frac{I_C}{\alpha}$$

$$V_E \approx 1/10^{th} \text{ of } V_{CC}$$

Voltage across R_C, $V_{RC} = V_{CC} - V_{CE} - V_E$

Equivalent resistance at the base of transistor, $R_B = \dfrac{(1+\beta)R_E}{10}$

Thevenin voltage at the base terminal, $V_{Th} = R_B I_B + V_{BE} + V_E$

$$R_1 = \frac{R_B V_{CC}}{V_{Th}} \quad \text{and} \quad R_2 = \frac{R_B V_{CC}}{V_{CC} - V_{Th}}$$

The frequency of oscillations

$$f_o = \frac{1}{2\pi\sqrt{(L_1 + L_2)C}}$$

Design Steps

Design of Biasing Circuit

Choose $V_{CC} = 2V_{CE} = 12 \text{ V}$

$$I_B = 0.017 \text{ mA}$$

$$\alpha = 0.994$$

$$I_E = 3.017 \text{ mA}$$

Taking emitter voltage as $1/10^{th}$ of V_{CC}, $V_E = 1.2$ V

$$R_E = 398\ \Omega$$

Voltage across R_C, $V_{RC} = 4.8$ V

$$R_C = 1.6\ \text{k}\Omega$$

Equivalent resistance at the base of transistor, $R_B = 7$ kΩ

Thevenin voltage at the base terminal, $V_{Th} = 2.02$ V

For the circuit to operate efficiently, it is assumed that the current through R_1 and R_2 is approximately equal and much larger than the base current.

$$R_1 = 41.58\ \text{k}\Omega$$

$$R_2 = 8.42\ \text{k}\Omega$$

Feedback Network Design

Taking $C = 1000$ pF, the value of $(L_1 + L_2) = 2.53$ mH

Choose $L_1 = 1$ mH and $L_2 = 1.53$ mH

In order to obtain a pure sine wave the value of R_E is adjusted using a 1 kΩ potentiometer in series with a fixed resistor of 100 Ω.

Choose $C_C = 10$ μF, 16 V and $C_E = 50$ μF, 16 V.

3.19 COLPITTS OSCILLATOR

Design Specifications

(*i*) Frequency of oscillations $f_O = 100$ kHz

(*ii*) The Q-point is to be located at $I_C = 3$ mA and $V_{CE} = 6$ V.

Use Si NPN transistors 2N2222 having $\beta_{min} = 100$ and $\beta_{typ} = 175$.

Theory

In the Colpitts oscillator circuit shown in Fig. 3.19, the oscillation frequency is determined by a tank circuit comprising of two capacitors and one inductor. The two capacitors are connected in series and the inductor is connected across them in parallel. The amplifier in *CE* configuration uses voltage-divider bias. In the circuit, C_C are the coupling capacitors and C_E is the emitter bypass capacitor. The phase-shift network consists of tank circuit formed by capacitors C_1, C_2 and inductor *L*.

The oscillations produced in the tank circuit are fed back to the base of transistor and appears in the amplified form across the collector and emitter of the transistor. The output voltage of the transistor will be in phase with the voltage across capacitor C_1. Since the junction of two capacitors is grounded, the voltage across C_2 that is actually fed back to the base will be 180° out of phase to that of the voltage across C_1, making the total phase shift in the circuit as 360°.

The frequency of oscillations is given by

$$f_o = \frac{1}{2\pi\sqrt{LC_{eq}}}$$

where $$C_{eq} = \frac{C_1 C_2}{C_1 + C_2}$$

Circuit Diagram

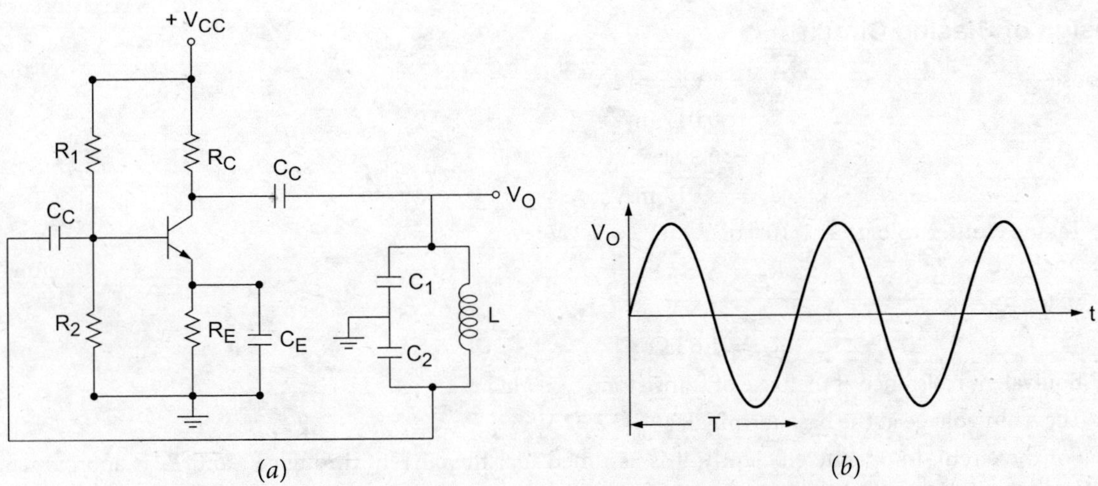

Fig. 3.19: (a) Colpitts oscillator using BJT and (b) output voltage waveform

Design Equations

$$V_{CC} \approx 2V_{CE}$$

$$I_B = \frac{I_C}{\beta}$$

$$\alpha = \frac{\beta}{1+\beta}$$

$$I_E = \frac{I_C}{\alpha}$$

$$V_E \approx 1/10^{th} \text{ of } V_{CC}$$

Voltage across R_C, $V_{RC} = V_{CC} - V_{CE} - V_E$

Equivalent resistance at the base of transistor, $R_B = \dfrac{(1+\beta)R_E}{10}$

Thevenin voltage at the base terminal, $V_{Th} = R_B I_B + V_{BE} + V_E$

$$R_1 = \frac{R_B V_{CC}}{V_{Th}}$$

$$R_2 = \frac{R_B V_{CC}}{V_{CC} - V_{Th}}$$

The frequency of oscillations is given by

$$f_o = \frac{1}{2\pi\sqrt{LC_{eq}}}$$

where $C_{eq} = \dfrac{C_1 C_2}{C_1 + C_2}$

Design Steps

Design of Biasing Circuit

Choose $V_{CC} = 2V_{CE} = 12$ V

$I_B = 0.017$ mA

$\alpha = 0.994$

$I_E = 3.017$ mA

Taking emitter voltage as $1/10^{th}$ of V_{CC}, $V_E = 1.2$ V

$R_E = 398\ \Omega$

Voltage across R_C, $V_{RC} = 4.8$ V

$R_C = 1.6$ kΩ

Equivalent resistance at the base of transistor, $R_B = 7$ kΩ

Thevenin voltage at the base terminal, $V_{Th} = 2.02$ V

For the circuit to operate efficiently, it is assumed that the current through R_1 and R_2 is approximately equal and much larger than the base current.

$R_1 = 41.58$ kΩ

$R_2 = 8.42$ kΩ

Feedback Network Design

Taking $L = 4$ mH, $C_{eq} = 633$ pF

Choose $C_1 = 1000$ pF. Then $C_2 = 1725$ pF

In order to obtain a pure sine wave the value of R_E is adjusted using a 1 kΩ potentiometer in series with a fixed resistor of 100 Ω.

Choose $C_C = 10$ μF, 16 V and $C_E = 50$ μF, 16 V.

3.20 CRYSTAL OSCILLATOR

Design Specifications

(*i*) Frequency of oscillations of the given crystal = 450 kHz

(*ii*) The Q-point is to be located at $I_C = 3$ mA and $V_{CE} = 6$ V.

Use Si NPN transistors 2N2222 having $\beta_{min} = 100$ and $\beta_{typ} = 175$. The crystal parameters are $L = 8.5$ H, $C_S = 0.015$ pF, $C_P = 5$ pF, $R = 1.05$ kΩ, and $Q = 23{,}200$.

Theory

The crystal oscillator using BJT is shown in Fig. 3.20. It employs a piezoelectric crystal, usually quartz. The resonant frequency and the quality factor Q depend upon the crystal dimensions, orientation of its surfaces with respect to its axes, and its mounting. Frequencies ranging from a few kilohertz to a few megahertz, and Q's in the range from several thousand to several hundred thousand, are commercially available. These extraordinarily high values of Q and the fact that the characteristics of quartz are extremely stable with respect to time and temperature account for the exceptional frequency stability of oscillators incorporating crystals.

To excite a crystal for operation in the series-resonant mode, it is connected as a series element in a feedback path as shown in the circuity. At the series-resonant frequency of the crystal, its impedance is small and the amount of positive feedback is large. Resistors R_1, R_2, and R_E provide a voltage divider stabilized dc bias. Capacitor C_E provides ac bypass of the emitter resistor and R_C is the collector resistance. The voltage fedback from collector to base is maximum when the crystal impedance is minimum. The resulting circuit frequency of oscillation is set by the series-resonant frequency of the crystal. Changes in supply voltage, transistor device parameters, etc have no effect on the circuit operating frequency, and is held stable by the crystal.

The series resonance frequency f_s of the crystal is

$$f_s = \frac{1}{2\pi\sqrt{LC_s}}$$

where L and C_S are the inductance and series capacitance of the crystal respectively.

The parallel resonance frequency f_p of the crystal is

$$f_p = \frac{1}{2\pi\sqrt{L\{C_S C_P/(C_S + C_P)\}}}$$

where C_P is the parallel capacitance of the crystal.

The frequency of oscillations f_O of the circuit is approximately equal to the series resonance frequency of the crystal.

$$f_o \approx f_s = \frac{1}{2\pi\sqrt{LC_s}}$$

Circuit Diagram

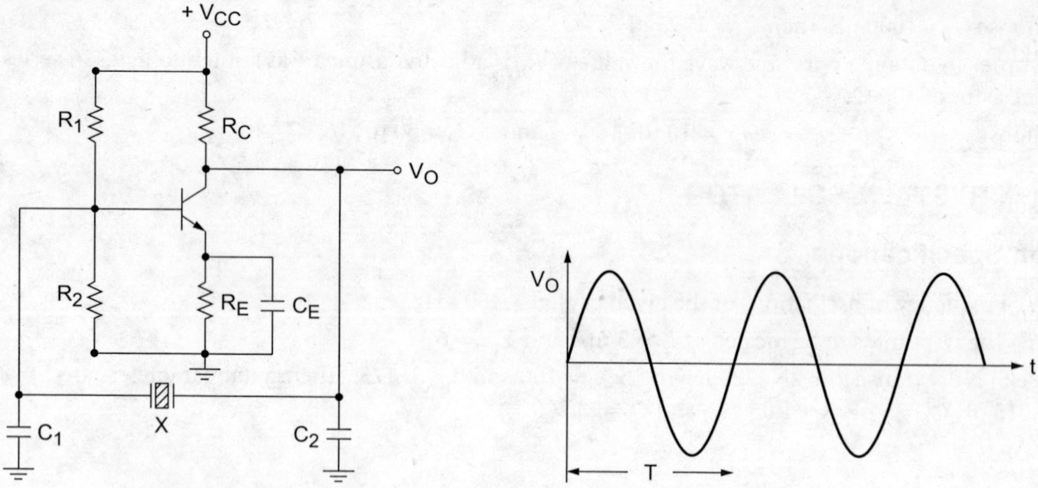

(a) (b)

Fig. 3.20: (a) Crystal oscillator using BJT and (b) output voltage waveform

Design Equations

$$V_{CC} \approx 2V_{CE}$$

$$I_B = \frac{I_C}{\beta}$$

$$\alpha = \frac{\beta}{1+\beta}$$

$$I_E = \frac{I_C}{\alpha}$$

$$V_E \approx 1/10^{\text{th}} \text{ of } V_{CC}$$

Voltage across R_C, $V_{RC} = V_{CC} - V_{CE} - V_E$

Equivalent resistance at the base of transistor, $R_B = \dfrac{(1+\beta)R_E}{10}$

Thevenin voltage at the base terminal, $V_{Th} = R_B I_B + V_{BE} + V_E$

$$R_1 = \frac{R_B V_{CC}}{V_{Th}}$$

$$R_2 = \frac{R_B V_{CC}}{V_{CC} - V_{Th}}$$

Design Steps

Design of Biasing Circuit

Choose $V_{CC} = 2V_{CE} = 12 \text{ V}$

$$I_B = 0.017 \text{ mA}$$

$$\alpha = 0.994$$

$$I_E = \frac{I_C}{\alpha} = 3.017 \text{ mA}$$

Taking emitter voltage as 1/10th of V_{CC}, $V_E = 1.2$ V

$$R_E = 398\ \Omega$$

Voltage across R_C, $V_{RC} = 4.8$ V

$$R_C = 1.6\ \text{k}\Omega$$

Equivalent resistance at the base of transistor, $R_B = 7$ kΩ

Thevenin voltage at the base terminal, $V_{Th} = 2.02$ V

For the circuit to operate efficiently, it is assumed that the current through R_1 and R_2 is approximately equal and much larger than the base current.

$$R_1 = 41.58\ \text{k}\Omega$$

$$R_2 = 8.42\ \text{k}\Omega$$

$$r_\pi = \frac{V_T}{I_B} = 1.52\ \text{k}\Omega$$

Choose $X_{C1} << r_\pi$ and the ratio of C_1 to C_2 between 1.5 and 2.

Let $$X_{C1} = \frac{r_\pi}{10} = 152\ \Omega$$

$$C_1 = 2.33 \text{ nF}$$

Taking $$\frac{C_1}{C_2} = 1.75, \quad C_2 = 1.33 \text{ nF}$$

In order to obtain a pure sine wave the value of R_E is adjusted using a 1 kΩ potentiometer in series with a fixed resistor of 100 Ω.

Choose $$C_E = 50\ \mu\text{F, } 16 \text{ V.}$$

3.21 SINGLE TUNED AMPLIFIER

Design Specifications

(*i*) Resonant frequency $f_r = 455$ kHz, the IF of an AM signal

(*ii*) The Q-point is to be located at $I_C = 3$ mA and $V_{CE} = 6$ V

(*iii*) Signal source resistance $R_s = 1$ kΩ

(*iv*) Load resistance $R_L = 10$ kΩ.

Use Si NPN transistors 2N2222 having $\beta_{min} = 100$ and $\beta_{typ} = 175$.

Theory

A single tuned amplifier in *CE* configuration is shown in Fig. 3.21(*a*). The tuned circuit formed by parallel *L* and *C* acts as a collector load and resonates at the frequency of operation as shown in Fig. 3.21(*b*). Resistors R_1, R_2, and R_E provides self-bias for the circuit. C_C are the coupling capacitors and C_E is the emitter bypass capacitor. The centre or resonance frequency of the circuit is given by,

$$f_r = \frac{1}{2\pi\sqrt{LC}}$$

Circuit Diagram

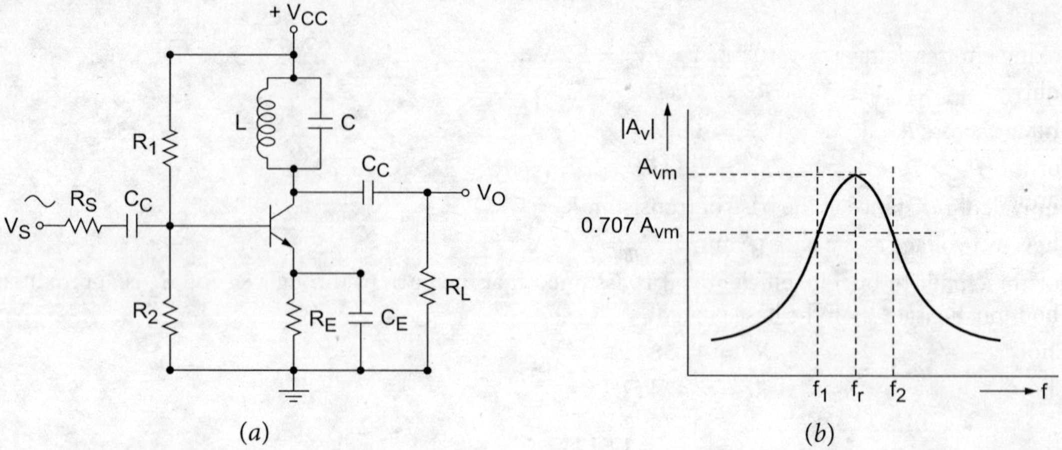

Fig. 3.21: (a) Single tuned amplifier using BJT and (b) frequency response characteristics

Design Equations

$$V_{CC} \approx 2V_{CE}$$

$$I_B = \frac{I_C}{\beta}$$

$$\alpha = \frac{\beta}{1+\beta}$$

$$I_E = \frac{I_C}{\alpha}$$

$$V_{E.} \approx 1/10^{th} \text{ of } V_{CC}$$

Equivalent resistance at the base of transistor, $R_B = \dfrac{(1+\beta)R_E}{10}$

Thevenin voltage at the base terminal, $V_{Th} = R_B I_B + V_{BE} + V_E$

$$R_1 = \frac{R_B V_{CC}}{V_{Th}}$$

$$R_2 = \frac{R_B V_{CC}}{V_{CC} - V_{Th}}$$

The resonant or centre frequency f_r is given by

$$f_r = \frac{1}{2\pi\sqrt{LC}}$$

Design Steps

Design of Biasing Circuit

Choose

$$V_{CC} = 2V_{CE} = 12 \text{ V}$$

$$I_B = 0.017 \text{ mA}$$

$$\alpha = 0.994$$
$$I_E = 3.017 \text{ mA}$$

Taking emitter voltage as 1/10th of V_{CC}, $V_E = 1.2$ V

$$R_E = 398\ \Omega$$

Equivalent resistance at the base of transistor, $R_B = 7$ kΩ

Thevenin voltage at the base terminal, $V_{Th} = 2.02$ V

For the circuit to operate efficiently, it is assumed that the current through R_1 and R_2 is approximately equal and much larger than the base current.

$$R_1 = 41.58 \text{ k}\Omega$$
$$R_2 = 8.42 \text{ k}\Omega$$

The IF frequency of an AM signal = 455 kHz. So, the resonant or centre frequency $f_r = 455$ kHz.

Choosing $L = 0.1$ mH, $C = 1.22$ nF

Choose $C_C = 10\ \mu$F, 16 V and $C_E = 50\ \mu$F, 16 V.

3.22 CLASS A POWER AMPLIFIER

Design Specifications

(*i*) Maximum power to be delivered to load $P_L(\text{max}) = 2$ W

(*ii*) Load resistance $R_L = 8\ \Omega$.

Theory

Figure 3.22 shows the basic *CE* amplifier without emitter resistor. The resistor R_B provides the biasing required to operate the circuit in class *A* mode. C_C couples the input signal V_s to the amplifier circuit.

The average DC power delivered by the supply voltage V_{CC} is

$$P_S = V_{CC} I_C$$

The average power delivered to load R_L or output power

$$P_L = \frac{V_m I_m}{2} = \frac{I_m^2 R_L}{2}$$

where V_m and I_m are the peak values of the AC output voltage and current, respectively. Using the minimum and maximum values of the output signals, V_m and I_m are expressed as

$$V_m = \frac{V_{CE}(\text{max}) - V_{CE}(\text{min})}{2}$$

$$I_m = \frac{I_C(\text{max}) - I_C(\text{min})}{2}$$

The output power, $$P_L = \frac{V_m I_m}{2} = \frac{[V_{CE}(\text{max}) - V_{CE}(\text{min})]\cdot[I_C(\text{max}) - I_C(\text{min})]}{8}$$

The maximum load power $P_L(\text{max})$ occurs when $V_{CE}(\text{min}) = 0$, $I_C(\text{min}) = 0$, $V_{CE}(\text{max}) = V_{CC}$, and $I_C(\text{max}) = 2I_C$. Thus

$$P_L(\text{max}) = \frac{V_{CC} 2I_C}{8} = \frac{V_{CC} I_C}{4}$$

The maximum conversion efficiency,

$$\eta_{max} = \left(\frac{P_{L\max}}{P_S}\right) \times 100\%$$

$$\eta_{max} = \left[\left(\frac{V_{CC}I_C}{4}\right) \Big/ (V_{CC}I_C)\right] \times 100\% = 25\%$$

The maximum collector dissipation is given by

$$P_{C\max} = \frac{V_{CC}I_C}{2}$$

The figure of merit, $$F_m = \frac{P_{C\max}}{P_{L\max}} = \left(\frac{V_{CC}I_C}{2}\right) \Big/ \left(\frac{V_{CC}I_C}{4}\right) = 2$$

Circuit Diagram

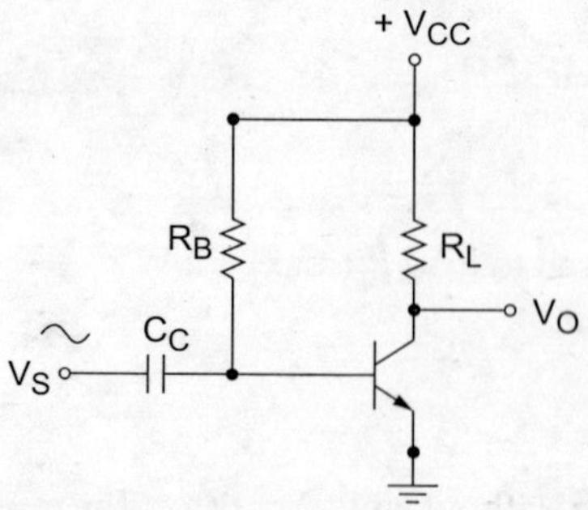

Fig. 3.22: Class A power amplifier using BJT

Design Equations

The maximum output signal power, $P_{L\max} = \dfrac{V_{CC}^2}{2R_L}$

The quiescent collector current, $I_{CQ} = \dfrac{V_{CC}}{R_L}$

The power supplied by V_{CC}, $P_S = V_{CC}I_{CQ}$

$$I_{BQ} = \frac{I_{CQ}}{\beta}$$

$$R_B = \frac{V_{CC} - V_{BE}}{I_{BQ}}$$

$$P_{RLdc} = I_{CQ}^2 R_L$$

Design Steps

$$V_{CC} = (2P_{L\max}R_L)^{1/2} = 5.7 \text{ V}$$

The quiescent collector current, $I_{CQ} = 0.7125$ A

The power supplied by V_{CC}, $P_S = 4.06$ W

Maximum collector dissipation = 4.06 W

A transistor having power dissipation of > 4.06 W is required. Choose 2N3019 NPN transistor having $I_{Cmax} = 1$ A, $P_{tot} = 5$ W, and $\beta = 50$

$$I_{BQ} = 0.01425 \text{ A}$$

$$R_B = 351\ \Omega$$

$$P_{RLdc} = 4.06 \text{ W}$$

Choose $$C_C = 10\ \mu\text{F},\ 10 \text{ V}$$

3.23 CLASS B COMPLEMENTARY-SYMMETRY POWER AMPLIFIER

Design Specifications

(*i*) Power to be delivered to load $P_L = 5$ W (*ii*) Load resistance $R_L = 4\ \Omega$.

Theory

Figure 3.23 shows the complementary-symmetry class *B* amplifier that does not use transformers. It uses a pair of complementary transistors, one NPN and other PNP. As transformers are not needed this makes the amplifier circuit much smaller for the same amount of output. Also, there are no stray magnetic effects or transformer distortion to affect the quality of the output signal. The resistors R_1 and R_2 are chosen such that the transistors are cut-off under no signal conditions and the circuit operates in class *B* mode. The capacitors C_C couples the input signal to the bases of two transistors and R_L is the load resistance. During positive half-cycle of the input, the NPN transistor Q_1 conducts and Q_2 is off. During negative half-cycle of the input, the PNP transistor Q_2 conducts and Q_1 is off.

While class *B* amplifiers have a much high gain than the class *A* types, one of the main disadvantages of class *B* type push-pull amplifiers is that they suffer from an effect known as crossover distortion. This occurs during the transition when the transistors are switching over from one to the other as each transistor does not stop or start conducting exactly at the zero cross-over point even if they are matched pairs. This is because the transistors require a base-emitter voltage greater than 0.7 V for them to start conducting, which results in both transistors being off at the same time.

The average input power supplied from the DC source is

$$P_S = V_{CC}I_{dc} = V_{CC}\left(\frac{2I_m}{\pi}\right) = 2V_{CC}\frac{I_m}{\pi}$$

The output or load power, $$P_L = \frac{I_m^2 R_L}{2} = \frac{V_m I_m}{2}$$

The conversion efficiency, $$\eta = \frac{P_L}{P_S}$$

$$\eta = \left(\frac{V_m I_m}{2}\right) \Big/ \left(\frac{2V_{CC}I_m}{\pi}\right) = \left(\frac{\pi}{4}\right)\left(\frac{V_m}{V_{CC}}\right)$$

For $V_m = \dfrac{2V_{CC}}{\pi}$, $$\eta = 50\%$$

The maximum power is delivered to load when $V_m = V_{CC}$ and is

$$P_L(\text{max}) = \frac{I_m^2 R_L}{2} = \frac{V_m I_m}{2} = \frac{V_{CC} I_m}{2} = \frac{V_{CC}^2}{2R_L}$$

The maximum power efficiency, $$\eta_{max} = \frac{P_L(\text{max})}{P_S} = \left(\frac{V_{CC}I_m}{2}\right) \Big/ \left(\frac{2V_{CC}I_m}{\pi}\right) = \frac{\pi}{4} \text{ or } 78.5\%$$

The average collector power dissipation for both transistors is given by

$$2P_C = P_S - P_L = \frac{2V_{CC}I_m}{\pi} - \frac{I_m^2 R_L}{2} = \frac{2V_m V_{CC}}{\pi R_L} - \frac{V_m^2}{2R_L}$$

Maximum collector dissipation occurs for $I_m(\max) = \dfrac{2V_{CC}}{\pi R_L}$ and the corresponding $V_m(\max) = I_m(\max)$ $R_L = \dfrac{2V_{CC}}{\pi}$.

The maximum collector dissipation, $2P_C(\max) = \dfrac{2V_{CC}^2}{\pi^2 R_L}$

The figure of merit, $$F_m = \frac{P_{C\max}}{P_{L\max}} = \left[\frac{V_{CC}^2}{(\pi^2 R_L)}\right] \bigg/ \left[\frac{V_{CC}^2}{2R_L}\right] = \frac{2}{\pi^2} \approx \frac{1}{5}$$

Circuit Diagram

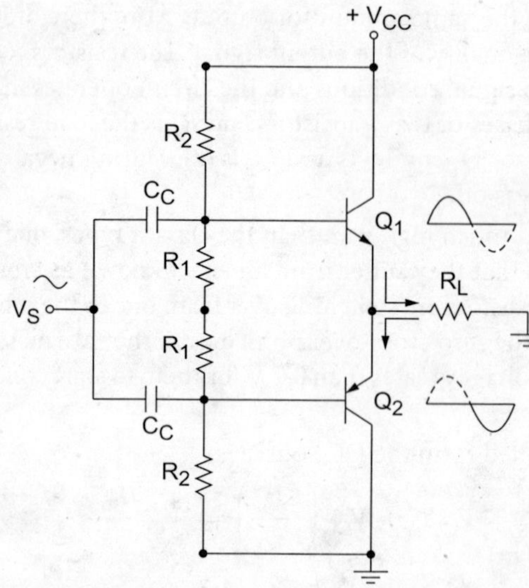

Fig. 3.23: Class B complementary-symmetry power amplifier using BJT

Design Equations

The output signal power, $P_L = \dfrac{V_{CC}^2}{2R_L}$

The maximum collector current, $i_{Cm} = \dfrac{2P_L}{V_{CC}}$

The average collector current, $I_C = \dfrac{2i_{Cm}}{\pi}$

Maximum power dissipation in the transistors, $P_{C\max} = \dfrac{2V_{CC}^2}{\pi^2 R_L}$

The average base current, $I_B = \dfrac{I_C}{\beta}$

The current I through R_1 and $R_2 >> I_B$, say, $10I_B$

$$V_{CC} = I(2R_1 + 2R_2)$$

For class B operation, the voltage drop across R_1 must be $< V_{BEcut\text{-}in}$ of transistor.

Design Steps

$$V_{CC} = (2P_L R_L)^{1/2} = 6.3 \text{ V}$$

The maximum collector current, $i_{Cm} = 1.587$ A

The average collector current, $I_C = 1.01$ A

Maximum power dissipation in the transistors, $P_{Cmax} = 2$ W

Maximum power dissipation in each transistor = 1 W

A transistor having collector current rating and power dissipation of at least 1.587 A and 1 W respectively is required. Choose BD131(NPN) and BD132(PNP) matched pair transistors, having $P_{Cmax} = 11$ W, $|I_{Cmax}| = 3$ A, and $\beta_{min} = 40$.

The average base current, $I_B = 25.25$ mA

Let the current I through R_1 and $R_2 = 10I_B = 0.2525$ A

$$V_{CC} = I(2R_1 + 2R_2)$$

$$V_{R1} < V_{BEcut\text{-}in}. \text{ Let } V_{R1} = 0.5 \text{ V}$$

$$6.3 = 1.0 + 2(0.2525)R_2$$

$$R_2 = 10.5\ \Omega$$

$$R_1 = \frac{0.5}{0.2525} = 1.98\ \Omega$$

Choose $C_C = 10\ \mu\text{F}, 10\text{ V}$

3.24 CLASS AB POWER AMPLIFIER

Design Specifications

(*i*) Power to be delivered to load $P_L = 5$ W (*ii*) Load resistance $R_L = 4\ \Omega$.

Theory

Figure 3.24 shows the complementary-symmetry class AB amplifier that minimizes or eliminates the cross-over distortion found in class B stage. It uses a pair of complementary transistors, one NPN and other PNP. The transistors operate in the active region when the input voltage is small (≈ 0 V). The transistors are biased in such a way that each transistor conducts for a small quiescent current I_Q without the input signal. During positive half-cycle of the input, the NPN transistor Q_1 conducts and during negative half-cycle of the input, the PNP transistor Q_2 conducts. The resistors R_1 and R_2 provide the quiescent current I_Q for the transistors and also ensure that the diodes D_1 and D_2 conduct. When diodes and transistors are of the same material, the thermal runaway is avoided. The capacitors C_C couples the input signal to the bases of two transistors and R_L is the load resistance.

The average power supplied from the DC source is

$$P_S = \frac{2V_{CC}I_m}{\pi} + V_{CC}I_Q = V_{CC}\left(\frac{2I_m}{\pi} + I_Q\right)$$

The output or load power, $P_L = \dfrac{I_m^2 R_L}{2} = \dfrac{V_m I_m}{2}$

The conversion efficiency, $\eta = \dfrac{P_L}{P_S}$

The maximum power is delivered to load when $V_m = V_{CC}$ and is

$$P_L(\max) = \frac{I_m^2 R_L}{2} = \frac{V_m I_m}{2} = \frac{V_{CC} I_m}{2} = \frac{V_{CC}^2}{2R_L}$$

The maximum power efficiency, $\eta_{\max} = \dfrac{P_L(\max)}{P_S}$

The collector power dissipation of both the transistors is

$$2P_C = P_S - P_{L\max} = V_{CC}\left(\frac{2I_m}{\pi} + I_Q\right) - \frac{I_m^2 R_L}{2}$$

Maximum collector dissipation occurs for $I_m(\max) = \dfrac{2V_{CC}}{\pi R_L}$ and the corresponding

$$V_m(\max) = I_m(\max) R_L = \frac{2V_{CC}}{\pi}$$

The maximum collector dissipation, $2P_C(\max) = \dfrac{2V_{CC}^2}{\pi^2 R_L} + V_{CC} I_Q$

The figure of merit, $F_m = \dfrac{P_{C\max}}{P_{L\max}} = \left[\dfrac{V_{CC}^2}{\pi^2 R_L}\right] \Big/ \left(\dfrac{V_{CC}^2}{2R_L}\right) = \dfrac{2}{\pi^2} \approx \dfrac{1}{5}$

Circuit Diagram

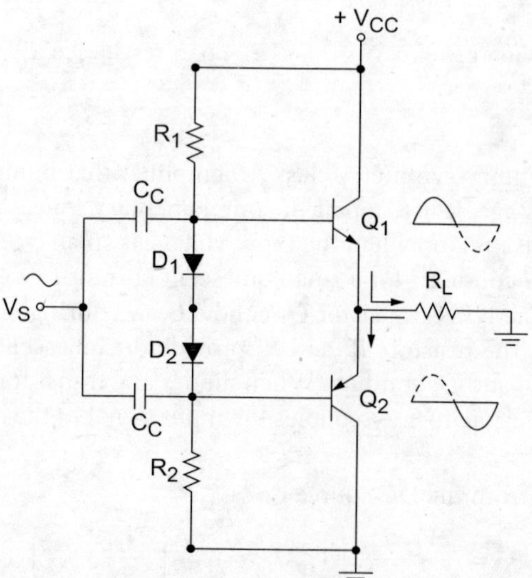

Fig. 3.24: Complementary-symmetry class AB power amplifier using BJT

Design Equations

The output signal power, $P_L = \dfrac{V_{CC}^2}{2R_L}$

The maximum collector current, $i_{Cm} = \dfrac{2P_L}{V_{CC}}$

The average collector current, $I_C = \dfrac{2i_{Cm}}{\pi}$

Maximum power dissipation in the transistors = $P_{Cmax} = \dfrac{2V_{CC}^2}{\pi^2 R_L}$

The average base current, $I_B = \dfrac{I_C}{\beta}$

The current I through R_1, R_2 and diodes >> I_B, say, $10I_B$

V_γ of diode is equal to $V_{BEcut\text{-}in}$ of transistor.

$$V_{CC} = I(R_1 + R_2) + 2V_\gamma$$

Design Steps

$$V_{CC} = (2P_L R_L)^{1/2} = 6.3 \text{ V}$$

The maximum collector current, $i_{Cm} = 1.587$ A

The average collector current, $I_C = 1.01$ A

Maximum power dissipation in the transistors = $P_{Cmax} = 2$ W

Maximum power dissipation in each transistor = 1 W

A transistor having collector current rating and power dissipation of at least 1.587 A and 1 W respectively is required. Choose BD131(NPN) and BD132(PNP) matched pair transistors, having $P_{Cmax} = 11$ W, $|I_{Cmax}| = 3$ A, and $\beta_{min} = 40$.

The average base current, $I_B = 25.25$ mA

Let the current I through R_1, R_2 and diodes = $10I_B = 0.2525$ A

For class *AB* operation, the diodes D_1 and D_2 and the transistors Q_1 and Q_2 must be of same material, so that V_γ of diode is equal to $V_{BEcut\text{-}in}$ of transistor. Choose 1N4148 diodes, which have a forward current of 0.3 A.

$$V_{CC} = I(R_1 + R_2) + 2V_\gamma$$

$$6.3 = 0.2525(R_1 + R_2) + 1.4$$

$$R_1 + R_2 = 19.4\ \Omega$$

Choose $R_1 = R_2 = 9.7\ \Omega \approx 10\ \Omega$

Choose $C_C = 10\ \mu\text{F},\ 10\text{ V}$

❑❑❑

4 CHAPTER CIRCUITS USING OPERATIONAL AMPLIFIERS AND LINEAR ICS

CIRCUITS USING OPERATIONAL AMPLIFIERS AND LINEAR ICS

Operational amplifiers or op-amps are linear devices and are one of the basic building blocks of analog electronic circuits. The op-amp is a voltage amplifying device and can perform a variety of different operations by making use of external feedback components such as resistors and capacitors, between its output and input terminals. These feedback components determine the resulting operation of the amplifier, giving rise to its name "operational amplifier". An operational amplifier consists of three terminals, two high impedance inputs and one output. One of the inputs is called the inverting input, marked with a "minus" sign (–). The other input is called the non-inverting input, marked with a "plus" sign (+). The output terminal can both sink and source either a voltage or a current. The op-amp requires a dual polarity power supply for its operation. The op-amp is basically a differential amplifier, as the output is proportional to the difference in voltage between the two inputs. An ideal op-amp is characterized by infinite open-loop gain, infinite input impedance, zero output impedance, infinite bandwidth, and zero offset voltage. Op-amps can be configured to perform a variety of operations such as amplification, addition, subtraction, differentiation, and integration. Op-amps finds application in virtually every area of analog and mixed analog-digital electronics.

The 555 timer IC makes use of three 5 kΩ resistors in its voltage divider network, hence the name 555 timer. This IC is useful for generating accurate time delays and oscillations. It is mainly useful for generating non-sinusoidal wave forms like square, ramp, pulse, etc. It consists of a voltage divider network, two comparators, one SR flip-flop, two transistors, and an inverter. The 555 timer is operated in monostable and astable modes to produce pulse and square wave outputs, respectively.

The 78xx voltage regulator ICs produce positive fixed DC voltage values, whereas, the 79xx voltage regulator ICs produce negative fixed DC voltage values. "xx" represents the magnitude of voltage that the voltage regulator IC provides. Both the ICs have 3 pins each, one pin is input, the second pin is ground, and the output is taken from the third pin. The capacitor Ci at the input terminals is used to prevent unwanted oscillations and the capacitor C0 at the output terminals acts as a line filter to improve the transient response.

The IC723 is generally used as series voltage regulator. It can be used to provide both positive and negative regulated voltages. It can provide up to 150 mA of current to the load, but this can be increased to more than 10 A by using power transistors. The input voltage ranges from 9.5 V to 40 V and it can regulate voltage from 2 V to 37 V. The working of IC can be explained by dividing it into two blocks, the reference voltage generator and the error amplifier. The reference voltage generator consists of a Zener diode. The error amplifier section consists of an error amplifier, a series pass transistor and a current limiting transistor. The error amplifier compares the output voltage applied at inverting input terminal through a feedback to the reference voltage applied at the non-inverting input terminal.

In this chapter, the design of various circuits using op-amps, 555 timer, 78xx, 79xx, and 723 ICs is given.

Note

(*i*) Unless otherwise specified, the op-amp used is 741, which has nominal supply voltage V of ±15 volts and slew rate of 0.5 V/μs.

(*ii*) The saturation voltages, V_{sat} = ±(V-2) volts = ±13 V

(*iii*) For higher slew rate applications, LM118/LM218/LM318 op-amps can be chosen, which have slew rate of 70 V/μs.

4.1 INVERTING AMPLIFIER

Design Specifications

(*i*) Voltage gain $A_V = 10$ (*ii*) Input resistance $R_{in} = 20$ kΩ.

Theory

The open loop gain A_{OL} of an operational amplifier is very high, as much as 10^6 (120 dB) or more. This very high gain makes the amplifier both unstable and hard to control as the input signals of the order of μV would be enough to cause the output voltage to saturate and swing towards one or the other of the supply voltage rails, losing complete control of the output. In order to reduce and control the overall gain of the amplifier, negative feedback is applied by connecting an external feedback resistor R_F between the output and the inverting input terminal.

The negative feedback results in the voltage at the inverting input terminal to be the sum of the input voltage plus the negative feedback voltage. The input signal is applied to the inverting terminal through an input resistor R_1. The non-inverting terminal is grounded. Since no current flows into the input terminals of the op-amp, the inverting terminal is at the same potential as the non-inverting terminal. A compensation resistor R_C is added as shown in the Fig. 4.1 to reduce the error in the output voltage caused by the voltage drops that result from the op-amp's input bias currents.

The closed-loop gain of the circuit is given by

$$A_v = -\frac{R_F}{R_1}$$

The negative sign in the equation indicates that the output voltage is 180° out of phase with the input signal. This is due to the feedback being negative. The maximum p-p output voltage swing, $V_{omax} = +V_{sat} - (-V_{sat}) = 2V_{sat}$.

The bandwidth of inverting amplifier is given by

$$BW = \frac{f_t R_1}{R_F + R_1}$$

where f_t is the unity-gain frequency. The product of DC gain and bandwidth is called as the gain-bandwidth product and indicates the frequency at which the differential gain drops to unity. This frequency is also called the unity-gain frequency. The slew rate is the highest rate of change possible in the output and is specified in volts per second. The slew rate limits the highest usable frequency of the op-amp. The highest sinewave frequency that can be amplified without slew rate distortion is given by

$$f_{SRL} = \frac{\text{Slew rate}}{\pi V_{omax}}$$

where f is the frequency of interest.

The input impedance of an op-amp is the impedance seen by the driving device.

Circuit Diagram

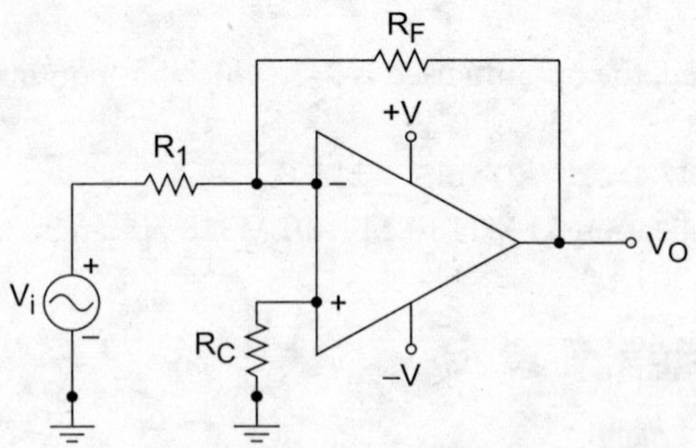

Fig. 4.1: Inverting amplifier

Design Equations

Voltage gain, $$A_v = -\frac{R_F}{R_1}$$

$$R_1 = R_{in} = \frac{V_i}{I_i}$$

$$R_F = R_1 \cdot |A_V|$$

Compensation Resistor $$R_C = R_1 \parallel R_F$$

Design Steps

$$R_1 = 20\ k\Omega$$
$$R_F = 200\ k\Omega$$
$$R_C = 18.18\ k\Omega$$

4.2 NON- INVERTING AMPLIFIER

Design Specifications

(*i*) Voltage gain $A_V = 32$ dB (*ii*) Input resistance $R_{in} = 50\ k\Omega$.

Theory

The non-inverting amplifier configuration is widely used to provide high input impedance. As shown in the circuit in Fig. 4.2, the input voltage is applied to the non-inverting terminal of the op-amp. The feedback is taken from the output of the op-amp via a resistor R_F to the inverting input of the operational amplifier where another resistor R_1 is taken to ground.

Since no current flows into the input terminals of the op-amp, the inverting and non-inverting terminals are at the same potential and the current flowing through the resistors R_1 and R_F is the same. The voltage gain of the non-inverting amplifier is given by

$$A_v = 1 + \frac{R_F}{R_1}$$

The positive sign in the equation indicates that the output voltage is in-phase with the input signal.

The maximum p-p output voltage swing, $V_{omax} = +V_{sat} - (-V_{sat}) = 2V_{sat}$. For supply voltages of ± V, the saturation voltages V_{sat} are typically ±(V-2) volts.

The bandwidth of inverting amplifier is given by

$$BW = \frac{f_t R_1}{R_F + R_1}$$

where f_t is the unity-gain frequency. The product of DC gain and bandwidth is called as the gain-bandwidth product and indicates the frequency at which the differential gain drops to unity. This frequency is also called the unity-gain frequency. The slew rate is the highest rate of change possible in the output and is specified in volts per second. The slew rate limits the highest usable frequency of the op-amp. The highest sinewave frequency that can be amplified without slew rate distortion is given by

$$f_{SRL} = \frac{\text{Slew rate}}{\pi V_{omax}}$$

$$V_{omax} = \frac{\text{Slew rate}}{\pi f}$$

where f is the frequency of interest.

Circuit Diagram

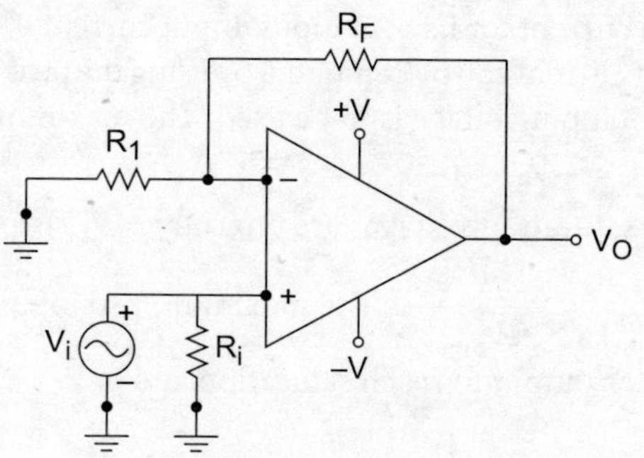

Fig. 4.2: Non-inverting amplifier

Design Equations

Voltage gain, $A_v = 1 + \frac{R_F}{R_1}$

Input resistance, $R_{in} = R_i$

Maximum p-p output voltage swing, $V_{omax} = +V_{sat} - (-V_{sat}) = 2V_{sat}$

Design Steps

Finding R_1 and R_F:

The resistors R_F and R_1 are chosen to produce the desired circuit voltage gain. If these resistors are chosen small, the feedback current becomes large. On the other hand, if R_F and R_1 values are chosen large, the non-ideal characteristics affects the circuit operation.

Given, $A_V = 32 \text{ dB} = 40.$

Choose R_1 as 1 kΩ.

$$R_F = R_1(A_V - 1) = 39\ \text{k}\Omega$$
$$R_i = R_{in} = 50\ \text{k}\Omega$$

4.3 INVERTING SUMMING AMPLIFIER

Design Specification

Output voltage $V_O = -(V_1 + 0.4V_2 + 5V_3)$ for the inputs V_1, V_2, and V_3.

Theory

The summing amplifier shown in Fig. 4.3 provides an output voltage V_o that is proportional to the algebraic sum of the input voltages, V_1, V_2, V_3, etc fed to the inverting terminal. This is because the input signals are effectively isolated from each other by the virtual ground node at the inverting terminal of the op-amp. The output voltage is given as

$$V_O = -\left[\left(\frac{R_F}{R_1}\right)V_1 + \left(\frac{R_F}{R_2}\right)V_2 + \left(\frac{R_F}{R_3}\right)V_3\right]$$

Voltage gain for each input signal, $A_v = -\dfrac{R_F}{R}$ $(R = R_1, R_2, R_3, \text{etc.})$

The input impedance Z_i seen by each input is equal to the value of the input resistor R on that particular input. Each source must supply the current for its own input. Input current, $I_i = V_i/R$. In the case of alternating voltage sources, the worst-case input current can be computed by using the peak values of input voltage. Similarly, for a variable DC source, maximum input voltage is to be used. The maximum p-p output voltage swing

$$V_{omax} = +V_{sat} - (-V_{sat}) = 2V_{sat}$$

For supply voltages of $\pm V$, the saturation voltages are typically $\pm$(V-2) volts.

Slew rate limiting frequency, $f_{SRL} = \dfrac{\text{Slew rate}}{\pi V_{omax}}$. The maximum p-p input voltage swing of an amplifier is the voltage that causes the amplifier's output to reach saturation and is $V_{imax} = \dfrac{V_{omax}}{A_v}$

The output impedance under closed-loop conditions is given as $\dfrac{\text{Output impedance (open-loop)}}{A_{OL}Z + 1}$

where A_{OL} is the open-loop gain of the op amp at a particular frequency and

$$Z = \frac{R_1 \,||\, R_2 \,||\, R_3}{(R_1 \,||\, R_2 \,||\, R_3) + R_F}$$

The worst-case value of A_{OL} occurs at the highest input frequency, *i.e.*, $A_{OL} = \dfrac{f_t}{f_i(\text{highest})}$.

The worst-case output or load current, $i_{Lmax} = \dfrac{V_{omax}}{R_L}$

Bandwidth, $BW = \dfrac{f_t R}{R_F + R}$, f_t is the unity-gain frequency. The bandwidth shall be computed for the input that gives largest closed-loop gain.

Maximum p-p output voltage swing, $V_{omax} = \dfrac{\text{Slew rate}}{\pi f}$, f is the frequency of interest.

The DC power supply voltages must be high enough to prevent saturation under worst-case input conditions. Generally, the condition to be considered is when all inputs are maximum at the same time.

The worst-case output will be

$$V_{Omax} = V_{O1max} + V_{O2max} + V_{O3max} = V_{1max}A_{V1} + V_{2max}A_{V2} + V_{3max}A_{V3}$$

Circuit Diagram

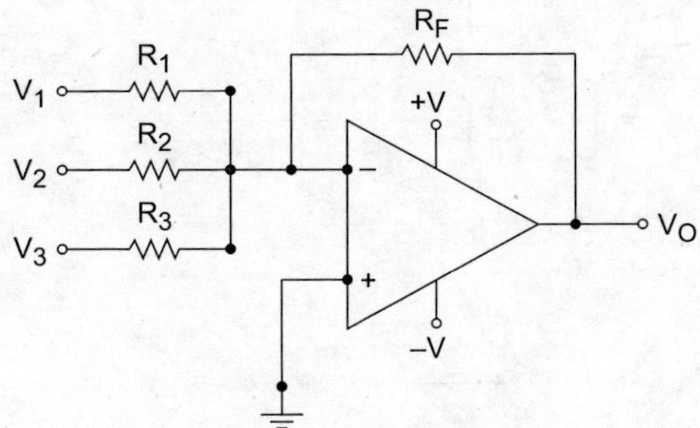

Fig. 4.3: Inverting summing amplifier

Design Equations

Voltage gain for each input signal, $A_v = -\frac{R_F}{R}$ $(R = R_1, R_2, R_3, \text{etc.})$

Input impedance, Z_i: The input impedance seen by each input is equal to the value of the input resistor R on that particular input.

Design Steps

Let

$$R_F = 10\ \text{k}\Omega$$

$$R_1 = -\frac{R_F}{A_{V1}} = 10\ \text{k}\Omega$$

$$R_2 = -\frac{R_F}{A_{V2}} = 25\ \text{k}\Omega$$

$$R_3 = -\frac{R_F}{A_{V3}} = 2\ \text{k}\Omega$$

4.4 NON-INVERTING ADDER

Design Specifications

(*i*) Inputs: $V_1 = 4 \sin 2000\ \pi t$, $V_2 = -0.5$ V DC, and $V_3 = +2$ V DC

(*ii*) Input impedance = 5 kΩ minimum.

Theory

Figure 4.4 shows the schematic diagram for a non-inverting adder circuit with three inputs. The input signals may be AC, DC, or some combination. The non-inverting adder generally exhibit superior high frequency performance when compared to the inverting type. Each input has an associated input resistor. The resistors isolate each source and prevents excessive source currents.

Unlike the inverting adder, the non-inverting adder does not take advantage of the virtual-ground summing node. The result is that individual channels will affect each other. The output voltage due to three inputs, V_1, V_2, and V_3 is given by

$$V_O = \left(1+\frac{R_F}{R_1}\right)\left[\frac{V_1(R_3\,||\,R_4)}{R_2+(R_3\,||\,R_4)}+\frac{V_2(R_2\,||\,R_4)}{R_3+(R_2\,||\,R_4)}+\frac{V_3(R_2\,||\,R_3)}{R_4+(R_2\,||\,R_3)}\right]$$

For $R_2 = R_3 = R_4 = R$, $$V_O = \left(1+\frac{R_F}{R_1}\right)\frac{(V_1+V_2+V_3)}{3}$$

In a more general case, $$V_O = \left(1+\frac{R_F}{R_1}\right)\frac{(V_1+V_2+V_3+\ldots+V_N)}{N}$$

where N is the number of inputs.

Circuit Diagram

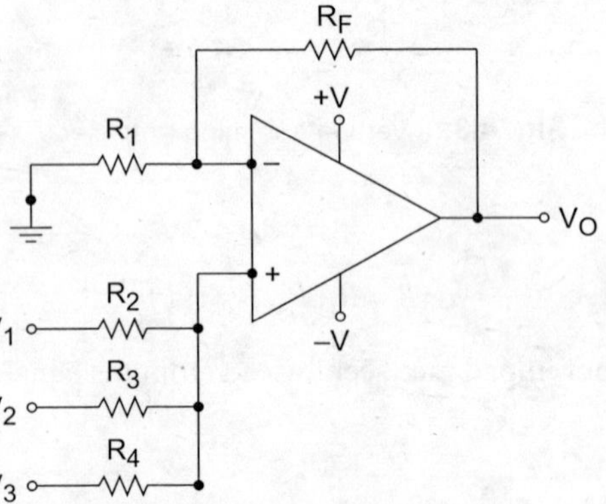

Fig. 4.4: Non-inverting adder

Design Equations

$$V_O = \left(1+\frac{R_F}{R_1}\right)\left[\frac{V_1(R_3\,||\,R_4)}{R_2+(R_3\,||\,R_4)}+\frac{V_2(R_2\,||\,R_4)}{R_3+(R_2\,||\,R_4)}+\frac{V_3(R_2\,||\,R_3)}{R_4+(R_2\,||\,R_3)}\right]$$

For $R_2 = R_3 = R_4 = R$, $$V_O = \left(1+\frac{R_F}{R_1}\right)\frac{V_1+V_2+V_3)}{3}$$

The input impedance, $$R_i = R\left\{\frac{1}{N-1}+1\right\}$$

where N is the number of inputs to be added.

Design Steps

Finding the values of R_1 to R_4:

Choose $R_1 = R_2 = R_3 = R_4 = R$

Resistor selection will determine the input impedance presented to the various signal sources. From the design requirements, the minimum value of R is found to be 3.33 kΩ.

Let us choose 3.9 kΩ value for R_1 to R_4.

Finding the feedback resistor, R_F: The value of the feedback resistor selected based on the requirement of voltage gain. Choosing a voltage gain equal to the number of inputs, *i.e.*, 3,

$$R_F = 7.8\text{ k}\Omega$$

The output voltage, $$V_O = \frac{3(4\sin 2000\pi t - 0.5 + 2)}{3} = 1.5 + 4\sin 2000\pi t \text{ volts}$$

Selection of the op-amp: For the addition of DC and AC signals, an op-amp with a low offset voltage, low drift, adequate bandwidth and slew rate is to be considered. Since in this case, the gain and bandwidth requirements are modest, let us choose a standard 741 op-amp.

4.5 INVERTING AC-COUPLED AMPLIFIER

Design Specifications

(*i*) Mid-band voltage gain, $A_{VM} = 10$

(*ii*) Lower cut-off frequency, $f_L = 200$ Hz

(*iii*) Input impedance, $Z_i \geq 2$ kΩ.

Theory

The AC-coupled inverting amplifier circuit is shown in Fig. 4.5. The input signal is coupled to the amplifier through the coupling capacitor *C*. The capacitor *C* blocks the dc components. At low frequencies, the coupling capacitor causes fall in the gain.

Since the inverting terminal of the op amp is at virtual ground, the output voltage is given by

$$V_O = -IR_F = -\left[\frac{V_i}{R_1 + 1/sC}\right]R_F$$

where *I* is the current through *C*, R_1 and R_F.

The voltage gain of the circuit is

$$A_V = \frac{V_O}{V_i} = -\frac{R_F}{R_1 + 1/sC}$$

The mid-band gain, $A_{VM} = -\frac{R_F}{R_1}$ and the lower cut-off frequency $f_L = \frac{1}{2\pi R_1 C}$.

Circuit Diagram

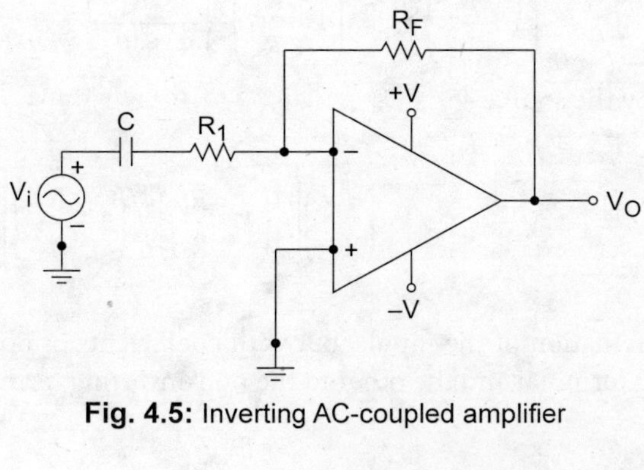

Fig. 4.5: Inverting AC-coupled amplifier

Design Equations

Since R_1 determines the input impedance of the amplifier, choose R_1 to meet the requirements of the input impedance.

$$R_F = R_1 \cdot |A_{VM}|$$

$$C = \frac{1}{2\pi f_L R_1}$$

Design Steps

Choose $R_1 = 2.7\ \text{k}\Omega$

$R_F = 27\ \text{k}\Omega$

$C = 0.295\ \mu\text{F}$

4.6 DIFFERENTIAL AMPLIFIER

Design Specifications

(*i*) Output voltage $V_O = 2V_2 - 3V_1$ (*ii*) Input impedance at each input = 50 kΩ.

Theory

The differential amplifier configuration shown in Fig. 4.6 has two inputs, V_1 and V_2, and one output V_O. It is a combination of the inverting and non-inverting voltage amplifiers. The output voltage is obtained by applying the superposition principle, as

$$V_O = V_{O1} + V_{O2}$$

where V_{O1} is the value of V_O with V_2 set to zero, and V_{O2} that with V_1 set to zero. For $V_2 = 0$, the circuit act as an inverting amplifier with respect to V_1.

So, $$V_{O1} = -\left(\frac{R_2}{R_1}\right)V_1$$

and the input resistance seen by the source V_1 is $R_{i1} = R_1$.

The circuit act as a non-inverting amplifier with respect to the voltage V_+ appearing at the non-inverting terminal when the input $V_1 = 0$. From the circuit,

$$V_+ = \frac{V_2 R_4}{R_3 + R_4}$$

Then, $$V_{O2} = \left(1+\frac{R_2}{R_1}\right)V_+ = \left(1+\frac{R_2}{R_1}\right)\cdot\left\{\frac{V_2 R_4}{R_3 + R_4}\right\}$$

The input resistance seen by the source V_2 is

$$R_{i2} = R_3 + R_4$$

$$V_O = V_{O1} + V_{O2} = \left(\frac{R_2}{R_1}\right)\left[\left\{\left(1+\frac{R_1}{R_2}\right)\Big/\left(1+\frac{R_3}{R_4}\right)\right\}V_2 - V_1\right]$$

The output resistance $R_O = 0$

The output is a linear combination of the inputs, but with coefficients of opposite polarity because one input is applied to the inverting terminal and the other to the non-inverting terminal of the op-amp.

When $\frac{R_3}{R_4} = \frac{R_1}{R_2}$, the resistances are said to form a balanced bridge and

$$V_O = \left(\frac{R_2}{R_1}\right)(V_2 - V_1)$$

The output is now proportional to the true difference of the inputs.

Circuit Diagram

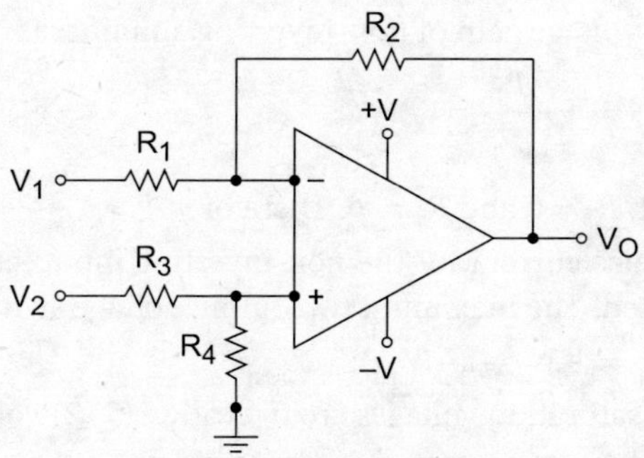

Fig. 4.6: Differential amplifier

Design Equations

$$V_O = V_{O1} + V_{O2}$$

$$V_{O1} = -\left(\frac{R_2}{R_1}\right)V_1 \quad \text{and} \quad V_{O2} = \left(1 + \frac{R_2}{R_1}\right)V_+ = \left(1 + \frac{R_2}{R_1}\right) \cdot \left\{\frac{V_2 R_4}{R_3 + R_4}\right\}$$

$$R_{i1} = R_1$$

$$R_{i2} = R_3 + R_4$$

Design Steps

Given,

$$R_{i1} = R_{i2} = 50 \text{ k}\Omega$$

$$R_1 = R_{i1} = 50 \text{ k}\Omega$$

$$R_3 + R_4 = 50 \text{ k}\Omega$$

Since,

$$V_O = 2V_2 - 3V_1$$

$$\frac{R_2}{R_1} = 3 \quad \text{or} \quad R_2 = 3R_1 = 150 \text{ k}\Omega$$

$$2 = 3\left[\left(1 + \frac{1}{3}\right) \Big/ \left(1 + \frac{R_3}{R_4}\right)\right] \Rightarrow \frac{R_3}{R_4} = 1$$

$$R_3 = R_4 = 25 \text{ k}\Omega$$

4.7 VOLTAGE FOLLOWER

Theory

Fig. 4.7 shows a unity gain buffer amplifier, constructed by applying a full series negative feedback to an op-amp simply by connecting its output to its inverting input, and connecting the signal source to the non-inverting input. In this configuration, the entire output voltage is placed contrary and in series with the input voltage. Thus, the two voltages are subtracted according to KVL and their difference is applied to the op-amp differential input. This connection forces the op-amp to adjust its output voltage simply equal to the input voltage, *i.e.*, V_o follows V_i, so the circuit is named op-amp voltage follower.

The input impedance of the circuit is very high, as the only current drawn from the source is the bias current for the non-inverting terminal. The output impedance, on the other hand, is quite low. The output impedance approaches an ideal value of 0, so the voltage follower circuit can interface a high impedance device or circuit to a lower impedance device or circuit. Although very little current is drawn from the source, a substantial current may be supplied to the load.

The general expression for the voltage gain of non-inverting amplifier is

$$A_v = 1 + \frac{R_F}{R_1}$$

For the voltage follower circuit, $R_F = 0$ and $R_1 = \infty$. Therefore, $A_v = 1$.

The input current is only the bias current for the non-inverting input terminal. This is ideally 0 and for most applications it may be neglected. The maximum p-p output voltage swing

$$V_{omax} = +V_{sat} - (-V_{sat}) = 2V_{sat}$$

For supply voltages of $\pm V$, the saturation voltages are typically $\pm(V\text{-}2)$ volts.

Slew rate limiting frequency, $f_{SRL} = \dfrac{\text{Slew rate}}{\pi V_{omax}}$.

Since the amplifier has a voltage gain of 1, the maximum input voltage swing is equal to the maximum output voltage swing. The output impedance of the voltage follower can be computed as follows.

$$R_o = \frac{\text{Output impedance (open loop)}}{A_{OL}}$$

where A_{OL} is the open-loop gain of the op-amp at the specified frequency and is given as

$$A_{OL} = \frac{f_t}{f_i}$$

where f_t is the unity-gain frequency and f_i is the specific input frequency being considered.

The bandwidth of the circuit is

$$BW = f_t$$

Circuit Diagram

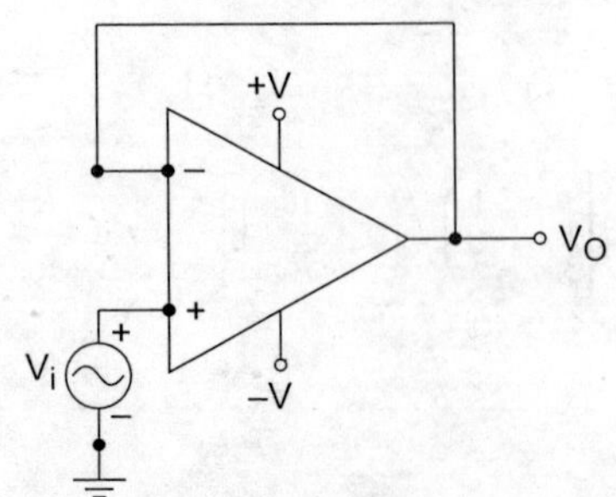

Fig. 4.7: Voltage follower circuit

Design Equations

For the non-inverting configuration, the voltage gain is

$$A_V = 1+\frac{R_F}{R_1}$$

where R_1 is the resistor connected to the inverting terminal and R_F is the feedback resistor.

For the circuit shown, $R_1 = \infty$ and $R_F = 0$. Therefore, $A_V = 1$.

Design Steps

$$R_1 = \infty \quad \text{and} \quad R_F = 0$$

4.8 INVERTING CURRENT AMPLIFIER

Design Specifications

(*i*) Current gain $A_i = -40$ (*ii*) Load resistance $R_L = 100\ \text{k}\Omega$.

Theory

The inverting current amplifier is shown in Fig. 4.8. Due to the parallel negative feedback connection at the input, the circuit input impedance is assumed to be zero. This implies that the inverting input terminal is at virtual ground. The current into the op-amp is negligible, so all input current flows through R_1 to node N. Effectively, R_1 and R_F are parallel.

Therefore, $V_{R1} = V_{RF}$

This means that a current is flowing through R_F, from ground to node N. These two currents combine to form the load current. The larger the I_{RF} is relative to I_i, more is the current gain.

The current gain, $A_i = -\frac{I_O}{I_i}$, $I_O = I_{R1} + I_{RF}$

$$I_{R1} = I_i \quad \text{and} \quad I_{RF} = \frac{V_{RF}}{R_F} = \frac{V_{R1}}{R_F} = \frac{I_i R_1}{R_F}$$

$$I_O = I_i\left(1+\frac{R_1}{R_F}\right) \quad \text{and} \quad A_i = -\left(1+\frac{R_1}{R_F}\right)$$

Circuit Diagram

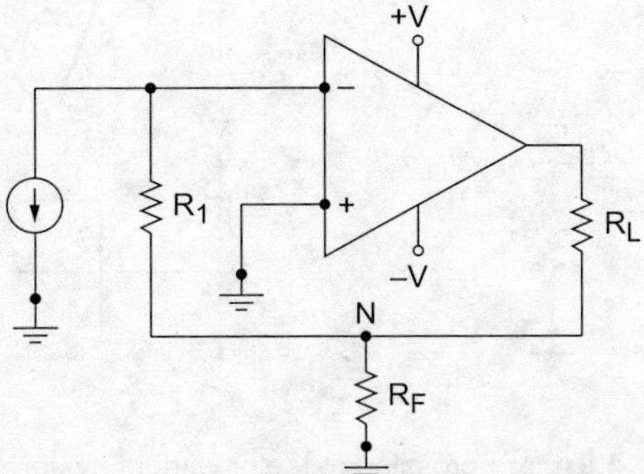

Fig. 4.8: Inverting current amplifier

Design Equations

$$A_i = -\frac{I_O}{I_i} \Rightarrow A_i = -\left(1+\frac{R_1}{R_F}\right) \Rightarrow \frac{R_1}{R_F} = -A_i - 1$$

Design Steps

$$\text{The maximum allowable input current} = \left(\frac{V_{sat}}{R_L}\right)\Big/A_i = \left(\frac{13\text{ V}}{100\text{ k}\Omega}\right)\Big/40 = 3.25\ \mu\text{A}$$

Let $R_F = 1\text{ k}\Omega.$

Then, $R_1 = 39\text{ k}\Omega$

4.9 ZERO-CROSSING DETECTOR WITHOUT HYSTERESIS

Design Specifications

(*i*) Input voltage $V_i = 0.4$ V peak (*ii*) Maximum input frequency = 15 kHz

(*iii*) Load resistance $R_L = 50\text{ k}\Omega$.

Theory

Figure 4.9(*a*) shows the schematic of a simple inverting voltage comparator being used as a zero-crossing detector. The output of the comparator switches every time the input signal passes through or crosses zero volts.

Consider the circuit to be an open-loop-amplifier. With no feedback, the gain of the amplifier is simply the open-loop gain of the op-amp itself. Since this gain value is very high, the output will be driven to either $+V_{sat}$ or $-V_{sat}$ if the input is more than perhaps 1 μV or so above or below ground potential.

Since the output voltage is at one of the two saturation levels at all times, except during the short switching time, the circuit essentially converts the sinewave input into a square wave. The resulting square wave shown in Fig. 4.9(*b*) will have the same frequency as the input, but the amplitude will always swing between $\pm V_{sat}$ regardless of the value of the input voltage.

Circuit Diagram and Waveforms

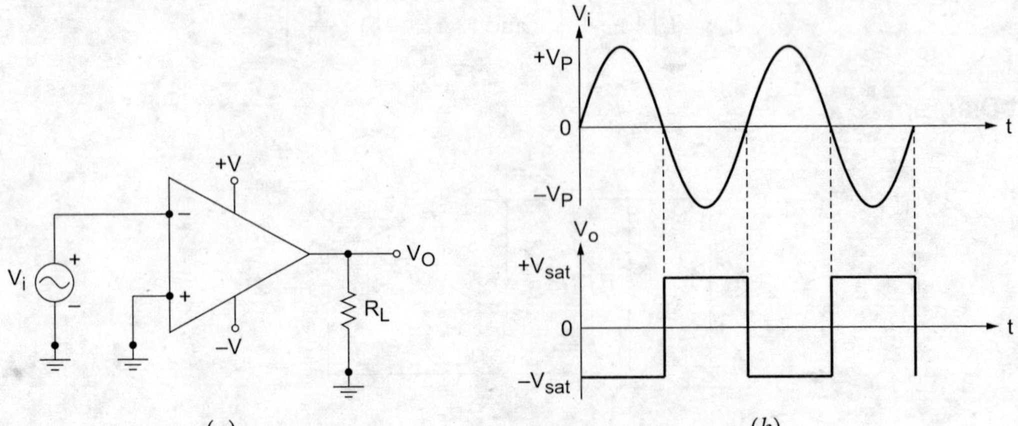

Fig. 4.9: Zero-crossing detector without hysteresis
(a) circuit diagram and (b) input and output voltage waveforms

Design Equations

The time for one one-half cycle, $t_1 = \frac{T}{2} = \frac{1}{2f}$

Assume the rise and fall times of no greater than 10 percent of the time for one-half cycle of the input signal, the maximum rise or fall times, $t_r(\text{max}) = t_f(\text{max}) = 0.1\ t_1$

The minimum acceptable slew rate for the op-amp = $\frac{[+V_{sat} - (-V_{sat})]}{t_r(\text{max})}$

Design Steps

Selection of op-amp: The op-amp that is chosen must satisfy the output voltage requirements, input voltage swings, and respond to the input frequencies.

The time for one one-half cycle is

$$t_1 = 33.33\ \mu\text{s}$$

For rise and fall times of no greater than 10 percent of the time for one-half cycle of the input signal, the maximum rise or fall times are

$$t_r(\text{max}) = t_f(\text{max}) = 3.333\ \mu\text{s}$$

The minimum acceptable slew rate for the op-amp is

$$\text{Slew rate (min)} = 7.8\ \text{V}/\mu\text{s}$$

Choose LM118 op-amp which satisfies the slew rate requirements.

4.10 ZERO-CROSSING DETECTOR WITH HYSTERESIS

Design Specifications

(*i*) Upper threshold = + 0.4 V

(*ii*) Lower threshold = – 0.4 V

(*iii*) Hysteresis = 0.8 V

(*iv*) Highest operating frequency = 20 kHz

(*v*) Maximum ratio of switching time to stable time, $\rho = 0.2$.

Theory

Fig. 4.10(*a*) shows the schematic diagram of a zero-crossing detector with hysteresis. It is a basic amplifier circuit in which the feedback is applied to the non-inverting input terminal. That is, the circuit is using positive feedback. Resistors R_F and R_1 form a voltage divider for the output voltage. The portion of the output voltage that appears across R_1 is fed to the non-inverting input of the op-amp. This voltage establishes the threshold voltage. When the output is positive, the voltage on the non-inverting input is called the positive or upper threshold voltage (V_{UT}). The voltage on this same input when the output is negative is called the lower or negative threshold voltage (V_{LT}). For the circuit shown, these two threshold levels will be having the same magnitude but opposite polarity.

Assume that the input voltage is at its most negative value and that the output is driven to its positive saturation level as shown in Fig. 4.10(*b*). A portion of the positive output voltage will be developed across R_1 and appear on the non-inverting input. This is the upper threshold voltage. As long as the input voltage is below the value of the upper threshold voltage, the circuit will remain in its present condition, *i.e.*, positive saturation.

Once the input voltage exceeds, *i.e.*, becomes more positive than the upper threshold voltage, the inverting terminal becomes more positive than the non-inverting terminal of the op-amp and the output will be driven all the way to its negative saturation limit. Once the output reaches its negative limit, the non-inverting input has a different voltage, the negative or lower threshold voltage on it. The circuit will remain in this stable condition as long as the input voltage is more positive than the lower threshold voltage.

When the input signal is rising, the switching point is determined by the upper threshold voltage. When the signal returns to a lower voltage, however, the output does not switch states as the upper threshold voltage is passed. Rather, the input voltage must go all the way down to below the lower threshold before the output will change states.

The difference between the two threshold voltages is called the hysteresis. The amount of hysteresis is determined by the values of the voltage divider consisting of R_F and R_1 and by the levels of output voltage, $\pm V_{sat}$.

Circuit Diagram and Waveforms

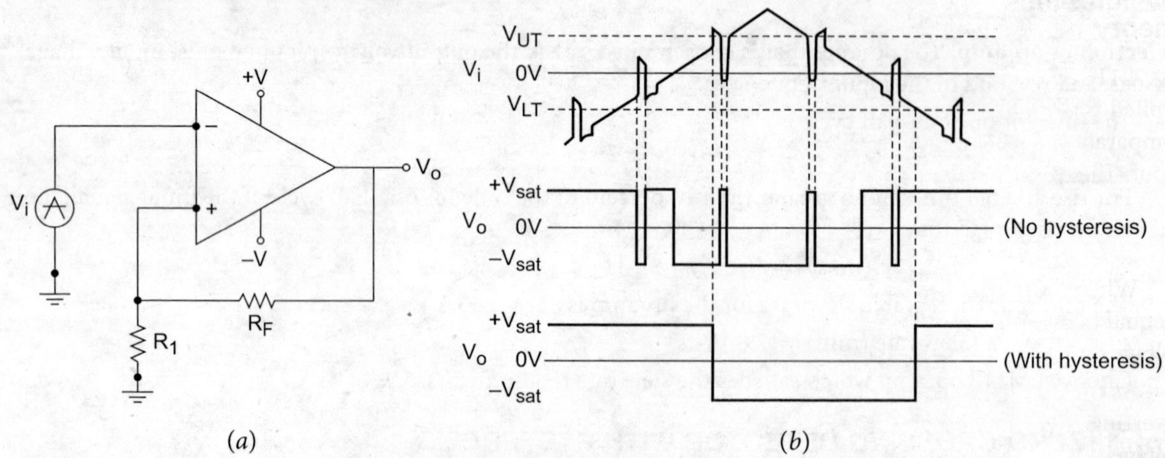

Fig. 4.10: Zero-crossing detector with hysteresis
(a) circuit diagram and (b) input and output voltage waveforms

Design Equations

Switching time $\quad t_s = \dfrac{\rho}{2(\rho+1)f_{max}}$

where ρ = maximum ratio of switching time to stable time.

$$\text{Slew rate (min)} = \frac{2(+V_{sat})}{t_s}$$

$$\frac{R_F}{R_1} = \left(\frac{2V_{sat}}{V_H}\right) - 1, \; V_H = \text{Hysteresis voltage}$$

Design Steps

Finding the required slew rate:

$$t_s = 4.17\ \mu s$$

$$\text{Slew rate (min)} = 6.235\ V/\mu s$$

Selection of op-amp: Choose LM118 op-amp which satisfies the slew rate requirements.

Finding R_F and R_1: The ratio of R_F and R_1 is dependent on the ratio of V_{sat} voltage to hysteresis voltage. The unloaded output swing will be approximately ± 13 V.

$$\frac{R_F}{R_1} = 31.5$$

Choosing $R_1 = 4.7\ k\Omega$, $R_F = 148\ k\Omega$.

4.11 VOLTAGE COMPARATOR WITH HYSTERESIS

Design Specifications

(*i*) Upper threshold voltage = + 2.5 V (*ii*) Lower threshold voltage = – 1.5 V

(*iii*) Highest operating frequency = 50 Hz.

Theory

Fig. 4.11(*a*) shows the schematic of an inverting voltage comparator with hysteresis. The input voltage V_i is applied to the inverting input of the op-amp. Resistors R_1 and R_2 form a voltage divider network across the comparator providing the positive feedback with part of the output voltage appearing at the non-inverting input. The amount of feedback β is determined by the resistive ratio of the two resistors used and is given as

$$\beta = \frac{R_1}{R_1 + R_2}$$

When the input signal is less than the reference voltage, *i.e.*, $V_i < V_{REF}$, the output voltage will be high and is equal to the positive saturation voltage, $+V_{sat}$. As the output is high and positive, the value of the reference voltage on the non-inverting input will be approximately equal to $+\beta V$ called the upper threshold (*UT*).

As the input signal V_i increases and becomes equal to the upper threshold point voltage V_{UT} at the non-inverting input, the comparator's output changes state, becomes low and equal to the negative saturation voltage, $-V_{sat}$. Now a second trip point voltage is created as a result of the negative saturation voltage at the output and a negative voltage $-\beta V_{sat}$ appears at the non-inverting input. The input signal must now fall below this second voltage level, called the lower threshold point (*LT*) for the voltage comparator's output to change or switch back to its original positive state.

When the output changes state, the reference voltage at the non-inverting input also changes creating two different reference voltage values and two different switching points, *UT* and *LT*. The difference between these two trip points is called hysteresis. Fig. 4.11(*b*) shows the hysteresis characteristics of the circuit. The amount of hysteresis is determined by the feedback fraction β of the output voltage fed back to the non-inverting input. For positive output voltages, $V_{REF} = +\beta V_{sat}$, but for negative output voltages, $V_{REF} = -\beta V_{sat}$. Then, the amount of voltage hysteresis, V_H is given as

$$V_H = V_{UT} - V_{LT} = +\beta V'_{sat} - (-\beta V_{sat}) = 2\beta V_{sat}$$

Circuit Diagram and Waveforms

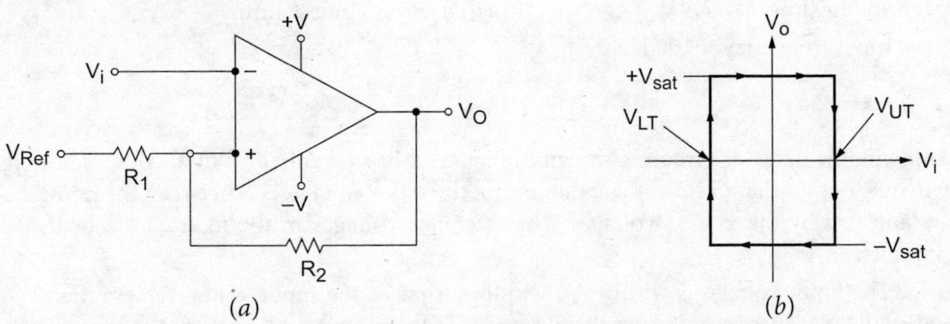

Fig. 4.11: Voltage comparator with hysteresis (a) circuit diagram and (b) hysteresis characteristics

Design Equations

$$\text{Slew rate(min)} = \pi f_{SRL} V_{omax} = \pi f_{SRL} 2V_{sat}$$

where V_{sat} is the op-amp saturation voltage and is typically $\pm(V\text{-}2)V$.

The feedback factor, $\beta = \dfrac{R_1}{R_1 + R_2}$

Upper threshold point voltage, $V_{UT} = \left[\dfrac{R_1}{R_1 + R_2}\right](+V_{sat}) = +\beta V_{sat}$

Lower threshold point voltage, $V_{LT} = \left[\dfrac{R_1}{R_1 + R_2}\right](-V_{sat}) = -\beta V_{sat}$

Hysteresis width, $V_H = V_{UT} - V_{LT} = +\beta V_{sat} - (-\beta V_{sat}) = 2\beta V_{sat}$

$$\frac{R_2}{R_1} = \left(\frac{2V_{sat}}{V_H}\right) - 1$$

The reference voltage, $V_{REF} = \dfrac{V_{LT} + V_H}{2}$

Design Steps

Finding the required slew rate:

$$\text{Slew rate(min)} = 0.0041\ \text{V/}\mu\text{s}$$

Selection of op-amp: The 741 op-amp satisfies the slew rate and the power supply requirements.

Finding R_1 and R_2: The ratio of R_2 and R_1 is dependent on the ratio of V_{sat} voltage to hysteresis voltage. For the lightly loaded ($R_L > 10$ kΩ) circuit, the output swing will be typically ± 13 V.

Hysteresis voltage, $V_H = 2.5 - (-1.5) = 4$ V

$$\frac{R_2}{R_1} = 5.5$$

Choosing, $R_1 = 33$ kΩ, $R_2 = 181.5$ kΩ.

$V_{REF} = 0.5$ V

4.12 WINDOW DETECTOR

Design Specifications

(*i*) Upper window limit = + 7.5 V

(*ii*) Lower window limit = + 5 V

(*iii*) Highest input frequency = 100 Hz.

Theory

A basic op-amp window detector circuit is shown in Fig. 4.12. It is essentially a dual comparator circuit and produces a two-state output that indicates whether or not the input voltage V_i is between the limits, *i.e.*, within the window established by the ± V_{REF} voltages. The reference voltages in the figure are established by two Zener diode circuits.

Assume that the input voltage is within the window. That is, the input voltage is less than + V_{REF} and greater than – V_{REF}. Under these conditions, the outputs of both op-amps will be driven to the + V_{sat} level. This reverse-biases the two isolation diodes D_3 and D_4 and allows the output V_O to rise to + 15 volts, indicating an "in window" condition.

If the input either exceeds + V_{REF} or falls below – V_{REF}, then the output of one of the two op-amps will go to – V_{sat} level and forward-bias its associated isolation diode. This will cause the output of the circuit to be pulled to –15 volts, ideally. In practice, the output voltage will be equal to the negative saturation level plus the forward voltage drop of the conducting isolation diode. This negative level indicates an "out of window" condition.

Circuit Diagram and Waveforms

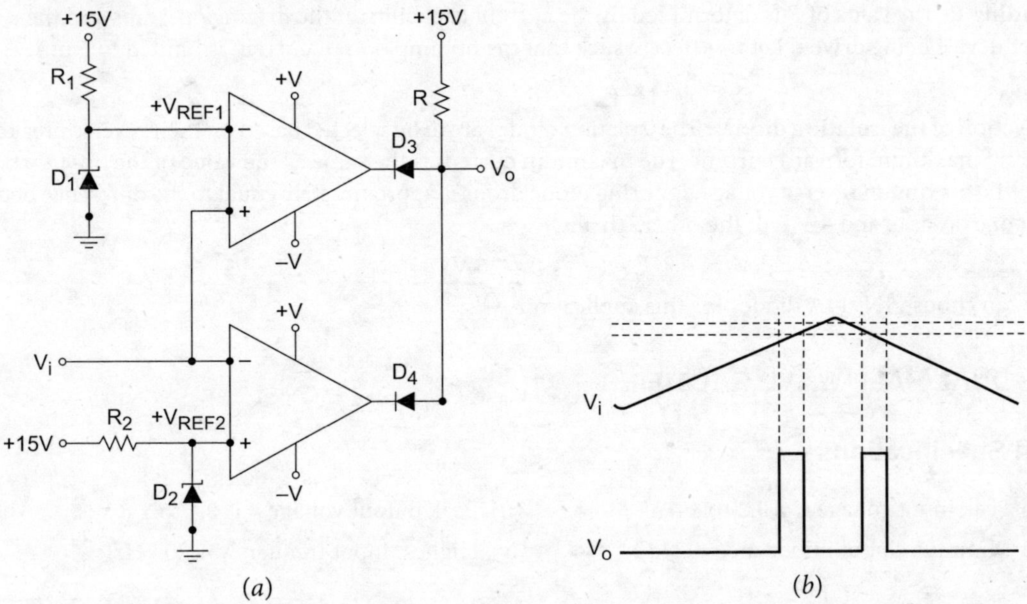

Fig. 4.12: Window detector (a) circuit diagram and (b) input and output voltage waveforms

Design Equations

Zener current limiting resistors are given by

$$R_1 = \frac{V_{CC} - V_{REF1}}{I_Z} \quad \text{and} \quad R_2 = \frac{V_{CC} - V_{REF2}}{I_Z}$$

where V_{CC} is the supply voltage, V_{REF1} is the lower window limit, I_Z is the Zener test current, and V_{REF2} is the upper window limit.

Pull-up resistor, $$R = \frac{V_{CC} - (-V_{sat}) - V_D}{I_R}$$

where V_{sat} is the saturation voltage of op-amp and is typically ±(V-2) volts, V_D is the forward voltage drop of the isolation diode and is typically 0.7 volts, and I_R is the current through R.

Ignoring diode drops, the peak inverse voltage of isolation diodes, $PIV = + V_{sat} - (- V_{sat})$.

Design Steps

Selection of the op-amp: Choose an op-amp like 741 which has better DC stability.

Selection of Zener diodes: From the data sheets of Zener diodes, the 1N5231 and 1N5236 devices will satisfy the requirements for the lower and upper reference voltages of 5 V and 7.5 V, respectively.

Finding the Zener current limiting resistors: From the data sheets, 20 mA is the test current for both diodes. The series resistor, taking Zener test current as the design value is

$$R_1 = 500\ \Omega \quad \text{and} \quad R_2 = 375\ \Omega$$

In this particular case, both Zener diodes are connected across the positive supply voltage because we require both references to be positive.

Finding *R*: The value of R is determined by the current capability of the driving op-amps and the type of circuit or device being driven. Let us select R such that the op-amp output current is limited to 5 mA.

$$R = 5.46\ \text{k}\Omega$$

Selection of the isolation diodes: The isolation diodes are to be selected based on their reverse breakdown voltage and maximum forward current. The maximum current is the same as the value of the op-amp output current. The maximum reverse voltage, ignoring diode drops is approximately equal to the difference between $+V_{sat}$ of one op-amp and $-V_{sat}$ of the other. That is,

$$PIV = +V_{sat} - (-V_{sat}) = 26\ \text{V}$$

We can choose 1N914A diodes for this application.

4.13 DUAL HALF-WAVE RECTIFIER

Design Specifications

(*i*) Peak input voltage = ± 0.2 to ± 4 V

(*ii*) Peak output voltage = ± 0.4 to ± 8 V

(*iii*) Minimum input impedance = 2 kΩ

(*iv*) Highest input frequency = 20 kHz.

Theory

The dual, half-wave rectifier circuit is shown in Fig. 4.13. Since the op-amp in figure is operated with closed-loop negative feedback, the inverting input pin is a virtual ground point. Thus, for practical purposes, one end of resistors R_1, R_2, and R_3 is connected to ground. This means that the total input voltage will be developed across R_1, causing an input current to flow through R_1. Since negligible current will flow in or out of the inverting terminal, all of the input current continues through either R_2 or R_3.

On positive input alternations, electron current will leave the output terminal of the op-amp and flow through D_1, R_3, and R_1 and out to the positive source. Since resistors R_1 and R_3 are equal, they will develop equal voltages, so we can expect the voltage across R_3 to be the same as the input voltage. Because one end of R_3 is grounded, the other end, V_{O1} provides a signal equal in amplitude to the positive alternation of the input signal. However, because of the direction of current flow through R_3, the polarity of the V_{O1} is inverted from V_i and produces negative half-wave waveforms.

During the negative alternation of the input cycle, electron current leaves the source and flows through R_1, R_2, and D_2 and into the output of the op-amp. Here again, because resistors R_1 and R_2 have the same current and are equal in value, they will have equal voltage drops. Thus, we expect a signal at V_{O2} having the same amplitude as the input. Additionally, as a result of the direction of the current flow, the V_{O2} output will provide a positive half-wave signal.

Circuit Diagram and Waveforms

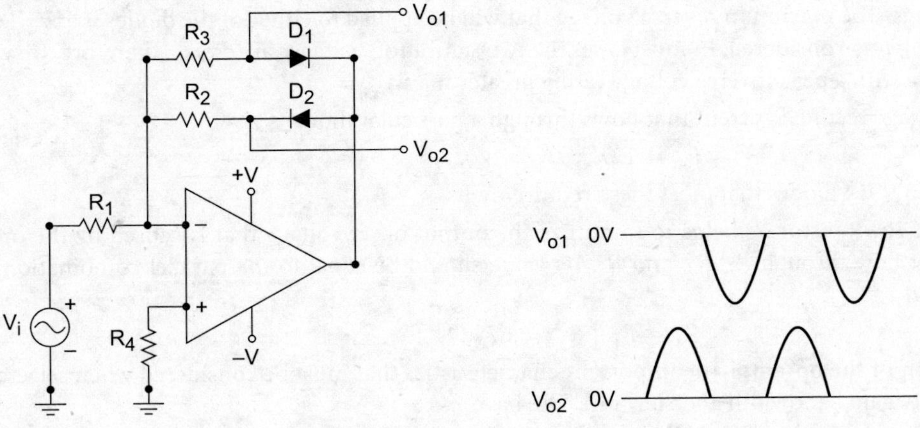

(a) (b)

Fig. 4.13: Dual half-wave rectifier (a) circuit diagram and (b) output voltages waveforms

Design Equations

Voltage gain,
$$A_V = \frac{V_O}{V_i}$$

$$A_V = -\frac{R_2}{R_1}$$

The average rectified current, $I_{av} = \dfrac{V_i(\text{peak})}{\pi R_1}$

$$A_{OL} = \frac{f_t}{f_i}$$

Slew rate for the op-amp $= \pi f_{SRL} V_O(\text{max})$

Design Steps

Finding the required voltage gain:

$$A_V = \frac{0.4}{0.2} = 2 \quad \text{or} \quad \frac{8}{4} = 2$$

By virtue of the circuit configuration, the gain will be inverting, *i.e.*, $A_v = -2$.

Finding R_1: The input impedance of an inverting amplifier is determined by the value of input resistor. In this case, resistor R_1 must be large enough to satisfy the minimum input impedance requirement. Select R_1 as 10 kΩ.

Finding R_2 and R_3: In the current design, resistors R_2 and R_3 will be equal because the gains for the two half cycles must be equal. We compute the value of R_2 or R_3 as

$$A_V = -\frac{R_2}{R_1} \quad \text{or} \quad R_2 = R_3 = -A_V R_1 = 20 \text{ k}\Omega$$

Selection of D_1 and D_2: The diodes are chosen based on the average forward current and peak inverse voltage ratings. The maximum reverse voltage that will be applied to either of the diodes is $\pm V_{sat}$, depending on the diode being considered. In most cases the two saturation voltages are equal. Therefore, we will need to select a diode with a peak inverse voltage rating greater than $V_{sat} = 13$ V.

The average rectified current that flows through a particular diode is given by

$$I_{av} = 127.3\ \mu\text{A}$$

Choose 1N914 diodes that meet these requirements.

Finding R_4: Resistor R_4 helps to minimize the output offset voltage that is caused by the op-amp bias currents that flow through R_1, R_2, and R_3. Its value should be equal to the parallel combination of R_1 and either R_2 or R_3.

$$R_4 = R_1 \,||\, R_2 = 6.67\ \text{k}\Omega$$

Selection of the op-amp: The important characteristics that must be considered when selecting the op-amp are unity gain bandwidth and slew rate. We have

$$A_{OL} = \frac{f_t}{f_i} \quad \text{or} \quad f_t = A_{OL} f_i = 40\ \text{kHz}$$

The required slew rate for the op-amp $= \pi(20 \times 10^3)15 = 0.94$ V/μs

So, choose LM118 op-amp which satisfies both the unity gain frequency and slew rate requirements.

4.14 IDEAL BIASED NEGATIVE SHUNT-CLIPPER

Design Specifications

(*i*) Clipping levels = + 2 to – 2 V

(*ii*) Input frequency = 2 kHz

(*iii*) Minimum input impedance = 10 kΩ.

Theory:

Figure 4.14 shows the schematic diagram of op-amp version of the biased shunt clipper circuit. Because the effects of the diode's forward voltage drop have been eliminated, the circuit performs like an ideal clipper circuit. Potentiometer *P* is used to establish the reference voltage or clipping level on the non-inverting input pin. Capacitor *C* prevents fluctuations in the clipping level.

For input voltages that are less than the value of the reference voltage, the inverting input is always less positive than the non-inverting input of the op-amp. Therefore, the output of the op-amp will remain at a positive level. The positive voltage on the output of the op-amp reverse-biases diode *D* and essentially opens the feedback loop. This allows the output of the op-amp to go to $+V_{sat}$ and remain there. With diode *D* reverse-biased, the op-amp circuit has no effect on the output signal.

We would expect V_o to be identical to V_i as long as we avoid excessive loading. For input signals that exceed the reference voltage, the output of the op-amp will start to move in the negative, *i.e.*, less positive direction. This causes diode *D* to become forward biased and closes the feedback loop. With a closed loop, we know that the voltage on the inverting input will be equal to the voltage on the non-inverting input, *i.e.*, $+V_{REF}$. Since the output is taken from this same point, it will therefore be equal to $+V_{REF}$ under these conditions. In other words, as long as the input voltage is more positive than $+V_{REF}$, diode *D* will remain forward biased, and the voltage on the inverting input and at the output of the circuit will remain at the $+V_{REF}$ level.

If diode *D* is reversed, then the circuit will clip the negative excursion of the input signal. Similarly, if the polarity of the reference voltage is reversed, the clipping will occur below 0.

Circuit Diagram and Waveforms

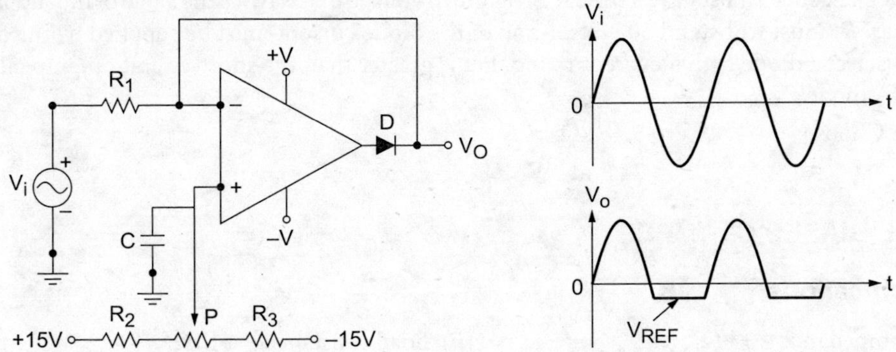

(a) (b)

Fig. 4.14: Ideal biased negative shunt-clipper (a) circuit diagram and (b) input and output voltage waveforms

Design Equations

The resistors R_2 and R_3 are given by

$$R_2 = \frac{P(V^+ - V_1)}{V_1 - V_2}$$

where V^+ is the + 15 V source, V_1 is the upper clipping level, and V_2 is the lower clipping level.

$$R_3 = \frac{P(V^- - V_2)}{V_2 - V_1}$$

where V^- is the – 15 V source.

$$\text{Slew rate} = 100 f_{max}[+V_{sat} - V_{REF}(\text{min})]$$

The capacitor $C = \dfrac{1}{2\pi f X_C}$ at the lowest operating frequency and $X_{C1} < 1/10^{th}$ of smaller of R_2 or R_3.

Design Steps

Finding the reference voltage divider: Since the design requires bipolar (± 2 V) clipping levels, we will use a potentiometer *P* as a voltage divider as shown in figure. If the potentiometer is too small, the power consumption is unnecessarily high. If the value is too high, then the non-ideal op-amp characteristics become more noticeable. Let us choose a 10 kΩ potentiometer.

The resistors R_2 and R_3 have (15 – 2) = 13 V across them. Additionally, they have the same current flow as *P*.

$$R_2 = 32.5 \text{ k}\Omega$$

Similarly,

$$R_3 = 32.5 \text{ k}\Omega$$

Finding R_1: The minimum value for R_1 is determined by the minimum required input impedance. The upper limit depends on the op-amp selected, desired output impedance, and required immunity to non-ideal characteristics. Let us select R_1 =12 kΩ.

Selection of the op-amp: The required slew rate can be estimated as

$$\text{Slew rate} = 3 \text{ V/}\mu\text{s}$$

This exceeds the 0.5 V/μs rating of the standard 741. So, choose LM118 op-amp which satisfies the slew rate requirements.

Selection of *D*: Diode *D* must have a reverse breakdown voltage that is twice the value of the supply voltage. In the present case, it must withstand 30 volts. As all of the diode current must be supplied by the op-amp, it is reasonable to select a diode with a current rating that is greater than the short-circuit op-amp current. Let us select 1N914A diode.

Finding C: Choose $X_C = 3.25\ \text{k}\Omega$

$$C = 24.5\ \text{nF}$$

4.15 IDEAL BIASED CLAMPER

Design Specifications

(*i*) Input impedance ≥ 1 kΩ (*ii*) Load resistance = 1 MΩ.

Theory

An ideal biased clamper circuit is shown in Fig. 4.15. It functions as though the diode were ideal *i.e.*, no forward voltage drop. This means that the circuit can be used to clamp signals in the millivolt range.

When the input reaches the negative peak, the output of the op-amp to go in a positive direction and will forward bias the diode *D*. The diode acts as a short circuit and the circuit behaves as a voltage follower with reference to the non-inverting input. The output of the op-amp will be at sum of the reference voltage and diode drop. The inverting input of the op-amp will also be at the reference voltage level. Capacitor *C* charges to the difference in potential between V_i and the reference voltage felt on the inverting input of the op-amp. The rate of charge for *C* is limited by the value of R_1. The potentiometer *P* fixes the reference voltage, V_{REF}.

For the remainder of the input cycle, the diode is reverse-biased. This effectively disconnects the op-amp from the circuit so the output will be same as the input plus the voltage across *C*. Again, C has no rapid discharge path, so it will act as a DC source and provide the clamping action.

In order for the clamper to be practical, it must drive into a very high impedance circuit. For this reason, the clamper circuit shown in figure is always followed by a voltage follower circuit. The discharge time constant should be much larger than the period of the input wave. The capacitor discharges through the load resistor. The discharge time constant is

$$T_{\text{disc}} = R_L C$$

Circuit Diagram and Waveforms

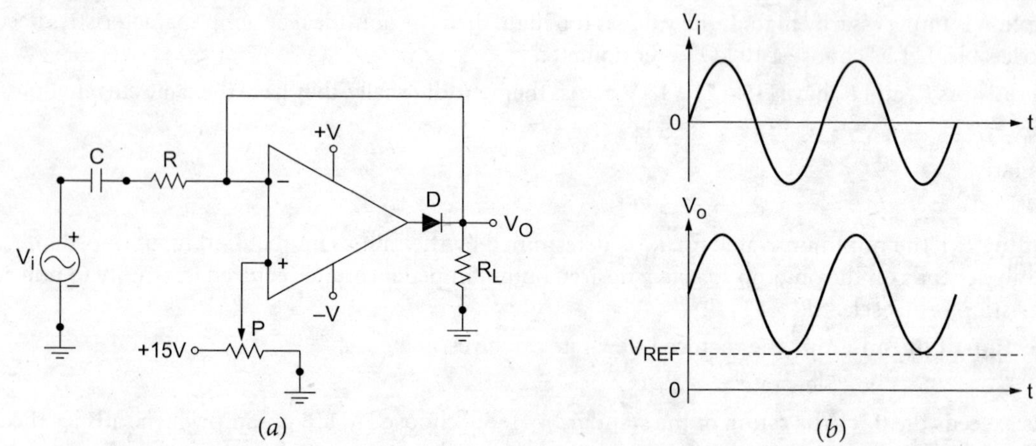

Fig. 4.15: Ideal biased clamper (a) circuit diagram and (b) input and output voltage waveforms

Design Equations

The input impedance of the circuit determines the resistor *R*.

$$R_L C >> T$$

where *T* is the period of the input waveform.

$$C = \frac{T_{\text{disc}}}{R_L}$$

Design Steps

Choose a 10 kΩ potentiometer *P*.

Since the input impedance is given as 1 kΩ, choose $R = 1.2$ kΩ.

Finding *C*:

For an input signal frequency of 1 kHz, $T = 1$ ms.

Let $$T_{\text{disc}} = 10T = 10 \text{ ms}$$

$$C = 10 \text{ nF}$$

Selection of *D*: Diode *D* must have a reverse breakdown voltage that is twice the value of the supply voltage. In the present case, it must withstand 30 V. The current rating for *D* is partially determined by the input impedance of the circuit being driven by the clamper. As all of the diode current must be supplied by the op-amp, it is reasonable to select a diode with a current rating that is greater than the short-circuit op-amp current. Let us select the 1N914A diode.

4.16 PEAK DETECTOR

Theory

Fig. 4.16 shows the schematic diagram of a peak detector circuit. The circuit is essentially an ideal clipper, followed by a parallel resistor and capacitor. The output of the clipper portion of the circuit will be a positive half-wave signal that is equal in amplitude to the peak of the input signal. Because *C* is connected to the same point, it will be charged to this peak voltage.

The time constant for charging *C* is very short and is equal to the product of *C* and the forward resistance R_f of the diode D_1. Thus, C charges almost instantly to the peak output of the clipper circuit. When *C* has charged to the peak input voltage and the output of op-amp swings to $-V_{sat}$ on the negative alternation, the peak inverse voltage of D_1 will be equal to the difference between $-V_{sat}$ and the maximum value of input voltage.

When the output of the clipper starts to decrease, diode D_1 becomes reverse-biased. This essentially isolates capacitor *C* and leaves the charge trapped. The only discharge path for *C* is through R_L. As this discharge time constant $R_L C$ is much longer than the charge time constant, *C* holds its charge.

Resistor R_L is the primary discharge path for *C*. If the input signal reduces its average amplitude, then *C* must be able to discharge to the new peak level. If the $R_L C$ time constant is too short, then the voltage on *C* will not be constant and will have a high value of ripple. On the other hand, if the $R_L C$ time constant is too long, then the circuit cannot respond quickly to changes in the input amplitude.

The diode D_2 conducts during negative half-cycle of V_i and prevents the op-amp from going into negative saturation. This further helps to lower the recovery time of the op-amp.

Circuit Diagram and Waveforms

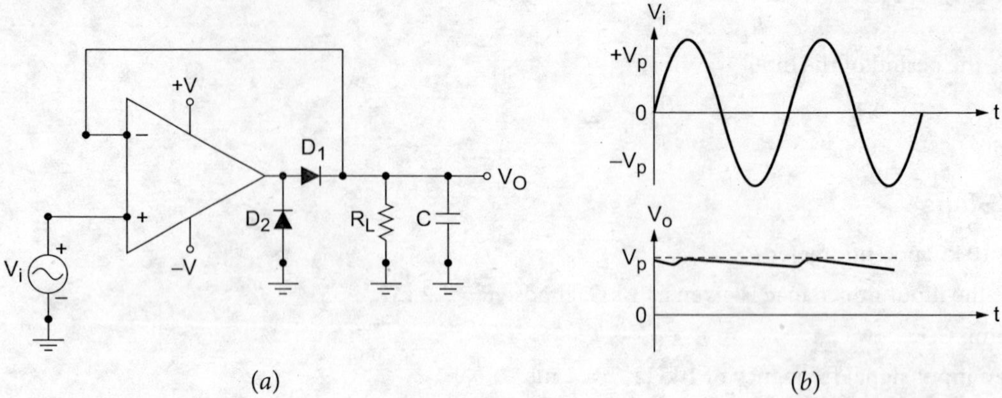

Fig. 4.16: Peak detector (a) circuit diagram and (b) input and output voltage waveforms

Design Equations

For proper operation of the circuit, the charging time constant $R_f C << T$ and discharging time constant $R_L C >> T$, T being the period of the input signal V_i.

The peak inverse voltage of the diode, $PIV \geq V_p - (-V_{sat})$

where V_p is the maximum or peak value of input voltage.

Design Steps

Selection of D_1: Assuming an input signal of amplitude 2 V and frequency 1 kHz, the peak inverse voltage of the diode is determined as

$$PIV = 2 - (-13) = 15 \text{ V}$$

The average current for D_1 is negligible, since it only conducts during charging of capacitor C and because C loses very little charge between consecutive cycles. However, the instantaneous current through D_1 can be considerably higher. The safest practice is to ensure that D_1 can survive the short-circuit current of op-amp. For the present design, choose 1N914A diodes for D_1 and D_2. This easily meets both the *PIV* and instantaneous current requirements.

Finding C: Let $$R_f C = \frac{T}{10}.$$

Assuming diode forward resistance of 100 Ω and for $T = 1$ ms, $C = 1$ μF.

Let $$R_L C = 100\,T$$

$$R_L = 100 \text{ k}\Omega$$

4.17 INTEGRATOR

Design Specifications

(*i*) Input frequency = 1 kHz minimum (*ii*) Unity gain frequency = 10 kHz.

Theory

The integrator is the circuit in which the output voltage waveform is the integral of the input voltage. Fig. 4.17 shows an integrator in the inverting configuration, consisting of a capacitor C_F in the feedback path. Resistors R and R_F determine the voltage gain of the circuit. The inverting input of the op-amp is a virtual ground point. A signal voltage V_i applied to the input terminal will cause a current to flow through resistor R and through R_F and C_F. Now, the voltage across R_F and C_F can only increase as fast as capacitor C_F can take on a charge. As the capacitor accumulates the charge, the current through R_F begins to change. The resistor R_F is chosen such that the current through it is much smaller than the capacitor current. As long as the input voltage is constant, the capacitor current is constant.

The expression for the output voltage V_o is given by

$$V_o = \frac{-1}{RC_F}\int_0^t V_i dt + C$$

where C is the integration constant and is proportional to the value of the output voltage V_o at time $t = 0$ seconds.

With a constant current through the capacitor, a linear ramp voltage is generated across it and therefore at the output of the op-amp. With a square wave input, the capacitor will periodically charge and discharge with equal, but opposite polarity currents. This produces a triangle wave at the output. Without R_F, the bias currents would charge C_F and cause an undesired DC offset at the output. This offset may even cause the amplifier to go into saturation. Resistor R_F is therefore included in the practical integrator circuit to reduce the gain at low frequencies.

The lower cut-off frequency f_1 and the unity-gain integration frequency f_0 are given as

$$f_1 = \frac{1}{2\pi R_F C_F}$$

$$f_0 = \frac{1}{2\pi R C_F}$$

Circuit Diagram and Waveforms

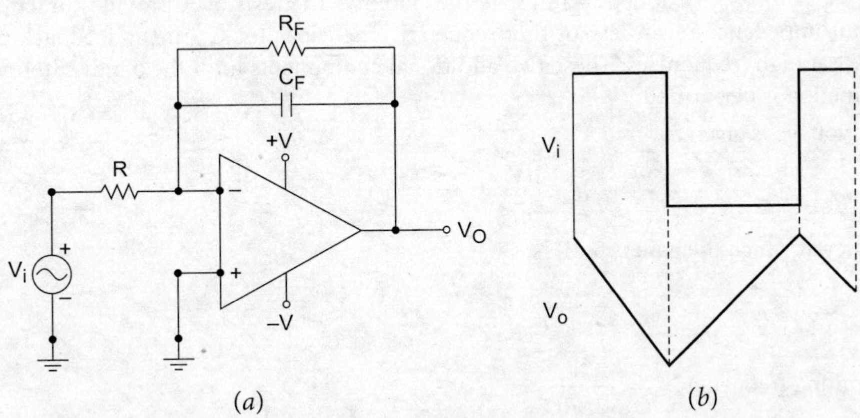

Fig. 4.17: Integrator (a) circuit diagram and (b) input and output voltage waveforms

Design Equations

The resistor R and the capacitor C_F are found to set the unity-gain integration frequency. The frequency f_1 at which the gain is limited is chosen about decade less than the minimum input signal frequency f_{min}.

$$f_1 = \frac{f_{min}}{10}$$

Design Steps

Finding R and C_F:

Let $$C_F = 0.01\ \mu\text{F}$$

$$R = \frac{1}{2\pi f_0 C_F} = 1.59\ \text{k}\Omega$$

Finding R_F:

$$f_1 = 100\ \text{Hz}$$

$$R_F = \frac{1}{2\pi f_1 C_F} = 159.15\ \text{k}\Omega$$

4.18 DIFFERENTIATOR

Design Specifications

(*i*) Input – square wave

(*ii*) Input frequency varies from 200 Hz to 2 kHz.

Theory

Figure 4.18 shows the schematic diagram of an op-amp differentiator circuit. The input signal is applied to the inverting input. The output of the circuit is differential of the input signal over a frequency range based on the circuit time constant and bandwidth of the amplifier. As the input impedance decreases with increasing frequency, the ideal differentiator circuit is inherently unstable and prone to oscillations. An input resistor R and a feedback capacitor C_F are added to improve the stability. The resistor R provides practical minimum limit to the input impedance regardless of the frequency. The capacitor C_F in the feedback path tends to bypass resistor R_F at high frequencies. These two additional components limit the bandwidth over which the differentiator function is performed.

The output voltage is given as

$$V_O = -R_F C \frac{dV_i}{dt}$$

The frequency at which the gain is 0 dB is

$$f_0 = \frac{1}{2\pi R_F C}$$

The gain limiting frequency, $f_1 = \dfrac{1}{2\pi RC}$

The values of f_1, RC, and $R_F C_F$ are selected such that $f_0 < f_1 < f_t$ where f_t is the unity gain frequency of op-amp.

Also, $$RC = R_F C_F$$

Proper differentiation of the input signal occurs when $R_F C \le T$, the time period of the input signal.

Circuit Diagram and Waveforms

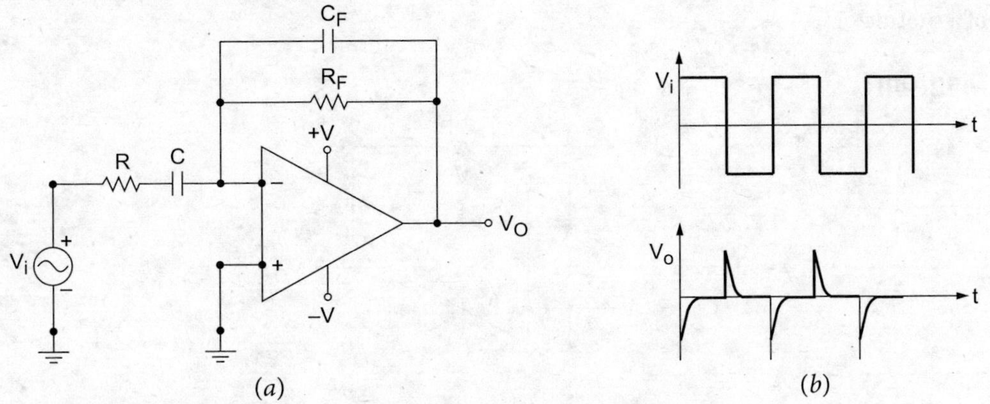

Fig. 4.18 Differentiator (a) circuit diagram and (b) input and output voltage waveforms

Design Equations

$$f_0 = \frac{1}{2\pi R_F C} \quad \text{and} \quad f_1 = \frac{1}{2\pi RC}$$

$$RC = R_F C_F$$

Design Steps

Finding C and R_F: Let f_0 = Maximum frequency of the input signal = 2 kHz

Assuming $C = 0.01$ μF, $R_F = 7.96$ kΩ

Finding R: Let $f_1 = 10f_0 = 20$ kHz. Then, $R = 796\ \Omega$

Finding C_F: Since $RC = R_F C_F$, $C_F = 1$ nF

4.19 SUBTRACTOR

Design Specifications

(*i*) Input voltages = 0 to + 5 V

(*ii*) Input frequency = 0 to 12 kHz

(*iii*) Input impedance > 3 kΩ.

Theory

Fig. 4.19 shows the schematic diagram of a basic subtractor circuit. If we assume that V_A is 0 volts, *i.e.*, grounded, the circuit functions as a basic inverting amplifier for input V_B. The voltage gain for this input will be determined by the ratio of resistors R_F and R_1. For $R_F = R_1$, the voltage gain is – 1, then the output voltage will be – V_B volts as a result of the V_B input signal.

When input V_B is grounded, the circuit functions as a basic non-inverting amplifier with respect to V_A. The overall voltage gain for the V_A input will be determined by the ratio of R_F and R_2 and the ratio of R_2 and R_3, which form a voltage divider on the input.

For $R_2 = R_3 = R_F$ the voltage at the non-inverting terminal = $\frac{V_A}{2}$ and the corresponding output voltage is V_A.

According to the superposition theorem, the output should be the net result of the two individual input signals. That is, the output voltage will be $+V_A - V_B$. Thus, we can see that the circuit does indeed perform the function of a subtractor.

Circuit Diagram

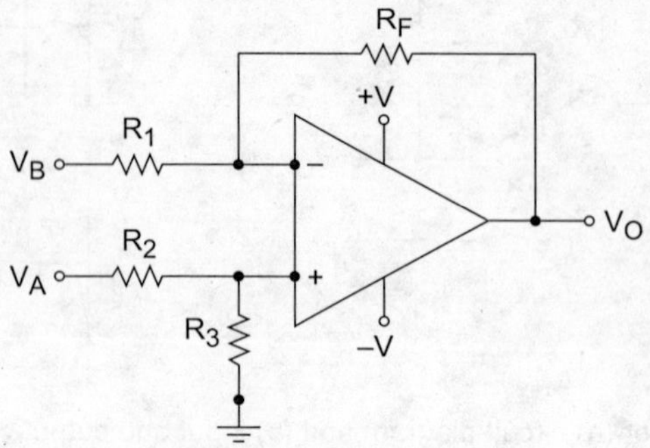

Fig. 4.19: Subtractor circuit

Design Equations

$$R = R_1 \text{ through } R_3 \text{ and } R_F = Z_i$$

$$\text{Slew rate (min)} = \pi f V_O(\text{max})$$

where V_O(max) is the maximum output voltage and is equal to $2\,V_{sat} = 26$ V.

The unity gain frequency required to meet the bandwidth requirements can be estimated as

$$f_t = BW \cdot A_V$$

where $A_V = 1 + \dfrac{R_F}{R_2}$ is the closed-loop gain of the non-inverting op-amp configuration = 2.

Design Steps

Finding R_1 through R_3 and R_F: The value of R_1 is determined by the minimum input impedance of the circuit.

$$R_1 = Z_i = 3\text{ k}\Omega$$

Resistors R_2, R_3 and R_F are set equal to R_1 in order to have the correct subtractor operation.

$$R_2 = R_3 = R_F = 3\text{ k}\Omega$$

Selection of the op-amp: The primary considerations for op-amp selection are slew rate and bandwidth. Unless the input signals are very small, the slew rate establishes the upper frequency limit.

$$\text{Slew rate (min)} = 0.98\text{ V}/\mu\text{s}$$

The unity gain frequency required to meet the bandwidth requirements can be estimated as

$$f_t = 24\text{ kHz}$$

Choose LM118 op-amp which satisfies both the unity-gain frequency and slew rate requirements.

4.20 ABSOLUTE VALUE CIRCUIT

Design Specifications

(*i*) Input voltage = $-10\text{ V} < V_i < 10\text{ V}$

(*ii*) Input impedance > 20 kΩ

(*iii*) Frequency range = 0 to 100 kHz.

Theory

The schematic diagram in Fig. 4.20 shows one way of obtaining the absolute value of a signal. The first stage of this circuit is a dual half-wave rectifier. For positive input signals, the output goes in a negative direction and forward-biases D_1. This completes the feedback loop through R_2. Additionally, the forward voltage drop of D_1 is essentially eliminated by the gain of the op-amp. The voltage at the junction of R_2 and D_1 will be the same magnitude but opposite polarity as the input voltage.

When a negative input voltage is applied to the dual half-wave rectifier circuit, the output of the op-amp goes in a positive direction. This forward-biases D_2 and completes the feedback loop through R_3. Diode D_1 is reverse-biased. In the case of basic dual half-wave rectifier circuit, the voltage at the junction of R_3 and D_2 is equal in magnitude but opposite in polarity to the input voltage. In the circuit shown, this voltage will be somewhat lower because of the loading effect of the current flowing through R_2, R_4, and R_5.

The outputs from the dual half-wave rectifier circuit are applied to the inputs of a difference amplifier circuit. Since the two half-wave signals are initially 180° out of phase, and since only one of them gets inverted by amplifier A_2, we can conclude that the two signals appear at the output of A_2 with the same polarity. In other words, both polarities of input signal produce the same polarity of output signal. By definition, this is an absolute value function.

Circuit Diagram and Waveforms

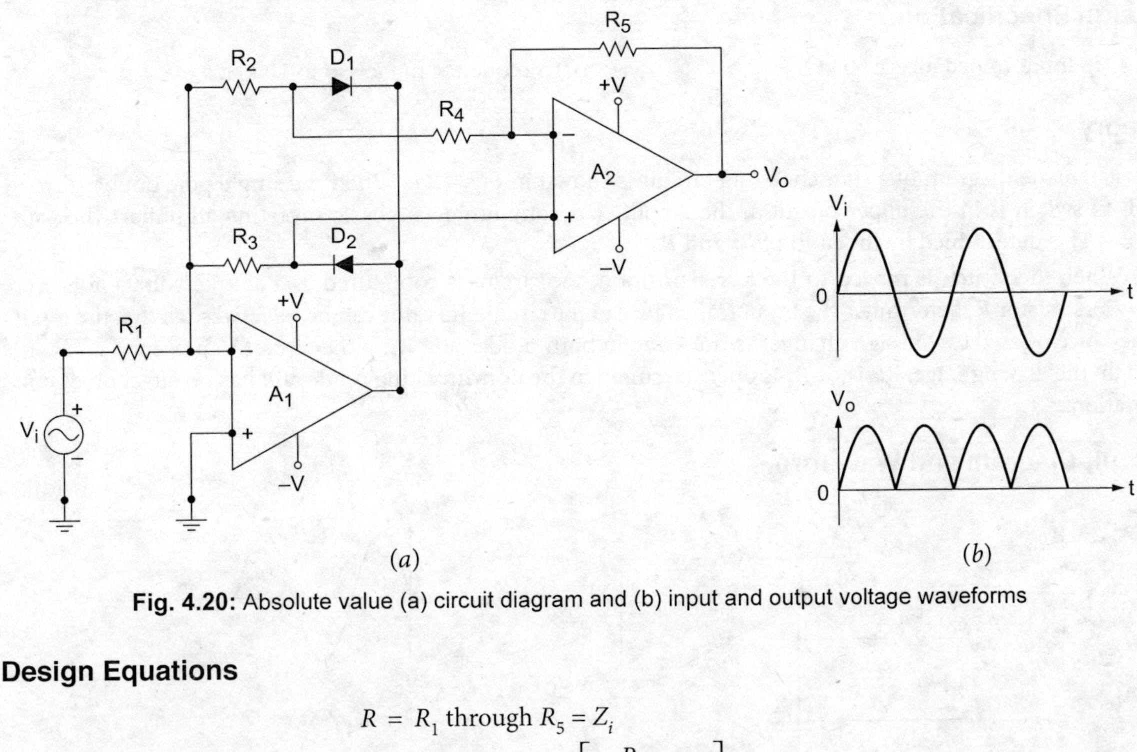

Fig. 4.20: Absolute value (a) circuit diagram and (b) input and output voltage waveforms

Design Equations

$$R = R_1 \text{ through } R_5 = Z_i$$

$$\text{Unity gain frequency for } A_2,\ f_t = BW \cdot A_V = BW\left[\frac{R_5}{R_2 + R_4} + 1\right]$$

$$\text{Slew rate (min)} = \pi f V_O(\text{max})$$

where $V_O(\text{max})$ is the maximum output voltage and is equal to $2V_{sat} = 26$ V

Design Steps

Finding the value of R (R_1 through R_5): The minimum value for all of the resistors is determined by the required input impedance. The maximum value is limited by the non-ideal characteristics of the circuit but is generally below 100 kΩ.

$$R \geq 20\ \text{k}\Omega$$

Let us choose 22 kΩ resistors for our design.

Selection of the op-amps: Slew rate and small signal bandwidth are used as the basis for op-amp selection. The required unity gain frequency for A_2 can be computed as

$$f_t = 150\ \text{kHz}$$

The bandwidth for A_1 will be somewhat lower because it has a voltage gain of – 1.

The required slew rate for either amplifier is

$$\text{Slew rate (min)} = 8.17\ \text{V}/\mu\text{s}$$

Choose LM118 op-amp which satisfies both the unity-gain frequency and slew rate requirements.

Choose 1N914A diodes for D_1 and D_2.

4.21 SIGN CHANGER

Design Specifications

(*i*) Input impedance ≥ 50 kΩ (*ii*) Frequency range = 0 to 10 kHz.

Theory

The schematic diagram of a sign changing circuit is shown in Fig. 4.21. When the single pole double-throw (SPDT) switch is in the upper position, the amplifier is configured as a basic inverting amplifier. The gain ($A_v = -1$) is determined by the ratio of R_1 and R_2.

When the switch is moved to the lower position, the circuit is configured as a simple voltage follower ($A_v = 1$). Resistor R_3 determines the input impedance of the circuit. Its value can be selected such that the input impedance offered by the sign changer is the same in both modes, allowing the circuit to present a constant load on the driving stage. Resistor R_1 is open-circuited in the non-inverting mode and has no effect on circuit operation.

Circuit Diagram and Waveforms

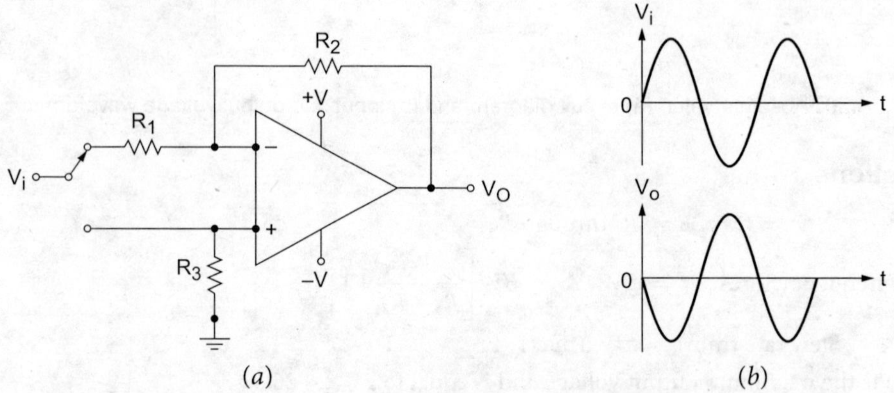

Fig. 4.21: (a) Sign changer circuit and (b) input and output voltage waveforms

Design Equations

$$R_1 = Z_i \quad \text{and} \quad R_2 = R_3 = R_1$$

The required unity gain frequency, $f_t = BW \cdot A_V$

$$\text{Slew rate} = \pi f V_O(\text{max})$$

where $V_O(\text{max})$ is the maximum output voltage and is equal to $2V_{\text{sat}} = 26$ V

Design Steps

Finding R_1 and R_2: The value of R_1 is established by the input impedance requirement.

$$R_1 = Z_i \geq 50 \text{ k}\Omega$$

Let us choose a standard value of 56 kΩ. Resistor R_2 will be the same value as R_1, since we want the circuit to have a voltage gain of – 1. $R_2 = 56$ kΩ.

Finding R_3: The value of R_3 should be equal to the value of R_1 so that we can maintain a constant input impedance for both modes. Therefore, $R_3 = R_1 = 56$ kΩ.

Selection of the op-amp: Slew rate and small signal bandwidth are used as the basis for op-amp selection. The required unity gain frequency can be computed as

$$f_t = 10 \text{ kHz}$$

The required slew rate for either amplifier is

$$\text{Slew rate} = 0.817 \text{ V/}\mu\text{s}$$

Choose LM118 op-amp which satisfies both the unity-gain frequency and slew rate requirements.

4.22 VOLTAGE REFERENCE CIRCUIT

Design Specifications

(*i*) Unregulated input voltage = + 10 to 15 V DC

(*ii*) Regulated output voltage = + 5 V

(*iii*) Reference current = 1 mA max.

Theory

Fig. 4.22 shows the circuit that combines a Zener diode and an op-amp to produce a simple but stable reference voltage. The input terminals of op-amp are essentially PN junctions connected to ground. This means that the voltage on either input will remain at 0.6V or less. Since the voltage across R is constant at approximately 0.6 V, its current is constant. It is essentially equal to the Zener diode current because the op-amp bias current is insignificant. Since the Zener current is constant, the Zener voltage will be constant.

Whenever the output voltage changes, this change is felt on the inverting terminal via D. Because the voltage on the inverting input is essentially limited by an internal junction, the changes fed back have only minimal effect on the voltage at the inverting terminal, but rather cause changes in bias currents. The result is that the output of the op-amp changes in a polarity that tends to compensate for the changing output voltage. Although the circuit compensates for changes in load current and in input voltage, it may still drift as a result of changes in temperature. This latter effect can be essentially eliminated by selecting a Zener diode with a temperature coefficient that is opposite (+ 2 mV/°C) to that in the op-amp.

The transistor Q is a simple current booster. The required Zener current may be substantially higher than the output current of the op-amp. Assuming that the junction breakdown voltages are adequate, the three critical parameters for the selection of Q are current gain β, power dissipation P_D, and collector current I_C.

The minimum required current gain for Q is given by,

$$\beta_{\text{min}} = \frac{I_{REF} + I_Z}{I_{\text{max}}}$$

where I_{max} is the maximum recommended output current for the op-amp.

The power dissipation of Q is,

$$P_D = (I_{REF} + I_Z)(V_{IN} - V_{REF})$$

Resistor R simply establishes the desired Zener current. It is given as

$$R = \frac{0.6}{I_Z}$$

Circuit Diagram

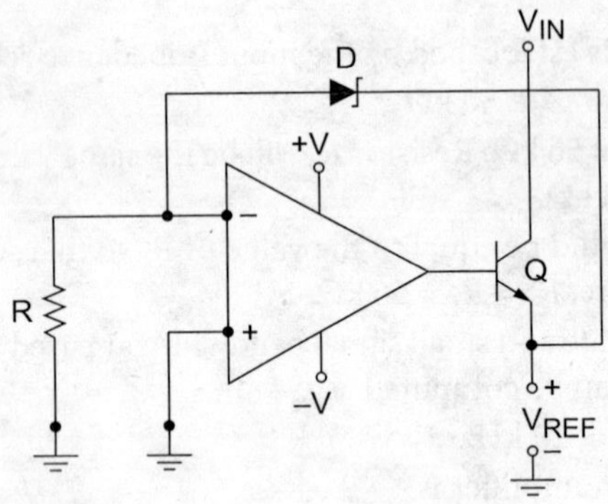

Fig. 4.22: Voltage reference circuit

Design Equations

The required breakdown voltage for the Zener diode D is $V_D = V_{REF} - 0.6$

$$R = \frac{0.6V}{I_{ZT}}$$

The minimum current gain of transistor, $\beta_{min} = \dfrac{I_{REF} + I_Z}{I_o}$

where I_{REF} is the maximum reference current, I_Z is the Zener current $= I_{ZT}$, and I_o is the maximum recommended output current for the op-amp considered.

The power dissipation of transistor, $P_D = (V_{IN} - V_{REF})(I_{REF} + I_Z)$

The collector current, $I_C = I_{REF} + I_Z$

Design Steps

Choosing *D*: The required breakdown voltage for the Zener diode D is given by

$$V_D = 4.4 \text{ V}$$

Referring to the manufacturer's data sheet, we can select 1N5229 Zener, which is rated for 4.3 V when a 20 mA current is passed through it. The value of R can be adjusted to obtain the required Zener voltage drop.

Finding *R*: The resistor R is given by

$$R = 30\ \Omega$$

Selection of *Q*: Because the highest DC input voltage given was + 15 V, the collector-to-emitter and collector-to-base breakdown voltages should be greater than 15 V.

The maximum recommended output current I_o for the op-amp LM118 considered = 4.5 mA.

$$\beta_{min} = 4.67$$

The power dissipation of Q is

$$P_D = 210 \text{ mW}$$

The collector current of Q, $I_C = 21$ mA

Let us choose 2N2222A as the current booster for the design.

4.23 SERIES VOLTAGE REGULATOR

Design Specifications

(*i*) Input voltage = + 12 to + 15 V (*ii*) Output voltage = + 5 to + 9 V

(*iii*) Output current = 0 to 0.5 A.

Theory

Figure 4.23 shows the schematic diagram of a series voltage regulator circuit. The input to the circuit is filtered but unregulated DC voltage and the output is regulated DC voltage that remains constant in spite of changes in the load current or changes in the input voltage. Transistor Q in figure is the series pass transistor.

The output voltage of the regulator circuit is sampled with the voltage divider made up of resistors R_1, R_2, and potentiometer P. The portion of the output that appears on the wiper arm of P is called the feedback voltage. Potentiometer P is used to adjust the amount of feedback voltage and thus is used to adjust the output voltage level. The voltage reference circuit provides a constant voltage level that can be used as a stable reference.

The op-amp in figure is called the error amplifier. It continuously compares the magnitude of the reference voltage with the level of the feedback signal, which represents the output voltage. Any difference between these two voltages, both magnitude and polarity, is amplified and applied to the base of the pass transistor. The polarity is such that the output voltage is returned to its correct value. If the positive voltage on the base of Q is decreased, it reduces the amount of forward bias and therefore increases the effective resistance of the pass transistor, causing an increased voltage drop across it and less drop across the load. Thus, the initial tendency for the load voltage to rise has been offset by an increased voltage drop across the pass transistor.

Circuit Diagram

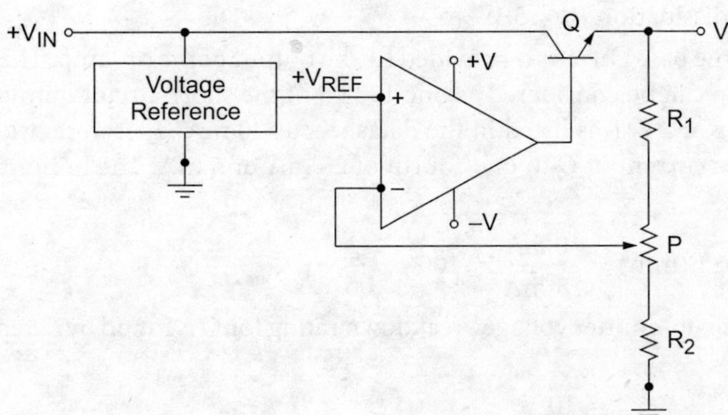

Fig. 4.23: Series voltage regulator circuit

Design Equations

Maximum collector current of transistor, $I_O(\text{max}) = \dfrac{P_D}{V_{IN}(\text{max}) - V_O(\text{min})}$

where P_D is the transistor power dissipation, V_{IN}(max) is the maximum input voltage, V_O(min) is the minimum output voltage.

$$P_D = I_O(\text{max})[V_{IN}(\text{max}) - V_O(\text{min})]$$

The minimum current gain β for the transistor is $\beta = \dfrac{I_C}{I_B}$

The maximum voltage across Q, $V_{CEO} = V_{IN}(\text{max}) - V_O(\text{min})$

The minimum and maximum voltage gains are

$$A_V(\text{min}) = \frac{V_O(\text{min})}{V_{REF}} \quad \text{and} \quad A_V(\text{max}) = \frac{V_O(\text{max})}{V_{REF}}$$

where V_{REF} is the reference voltage.

The minimum value for the potentiometer *P* is chosen much larger than the minimum equivalent load resistance.

The minimum voltage gain of a non-inverting amplifier in the circuit is,

$$A_V(\text{min}) = \left[\frac{R_1}{R_2 + P} + 1\right]$$

The maximum voltage gain of a non-inverting amplifier in the circuit is

$$A_V(\text{max}) = \left[\frac{R_1 + P}{R_2} + 1\right]$$

The resistor $R_1 = [A_V(\text{min}) - 1](R_2 + P)$ and $R_2 = \dfrac{PA_V(\text{min})}{A_V(\text{max}) - A_V(\text{min})}$

Design Steps

Selection of the pass transistor: The characteristics of the pass transistor are determined by the input voltage, the output voltage and current requirement, and the output drive capability of the op-amp. The collector current rating of the transistor must be greater than the value of load current. This means that the transistor must have a maximum DC current rating of greater than 0.5 A.

The transistor power dissipation $P_D = 5$ W

In the circuit shown, the base current is provided by the output of the op-amp. The maximum current to be provided by the op-amp can be considered as one-fourth of the short-circuit output current rating of the op-amp, which, in the case of a 741, is listed in the datasheet as 20 mA. Therefore, we shall limit the output current and hence the base current of Q to one-fourth of 20 mA or 5 mA. The minimum current gain for Q can now be determined as

$$\beta(\text{min}) = \frac{0.5\text{ A}}{5\text{ mA}} = 100$$

The minimum collector-to-emitter voltage breakdown rating for Q is found by determining the maximum voltage across Q. That is,

$$V_{CEO} = 10\text{ V}$$

Let us select BD131 transistor for this application.

Finding the required op-amp voltage gain: The minimum and maximum voltage gains, assuming reference voltage of + 4 V, are

$$A_V(\text{min}) = 1.25 \quad \text{and} \quad A_V(\text{max}) = 2.25$$

Finding the value for P: Let $P(\min) = 10R_L$

The minimum equivalent load resistance, $R_L = \dfrac{5\text{ V}}{0.5\text{ A}} = 10\ \Omega$

$$P(\min) = 100\ \Omega$$

The maximum value can be few tens of thousands of ohms. Let us choose a 5 kΩ potentiometer for *P*.

Finding R_1 and R_2: $R_2 = 6.25\text{ k}\Omega$

$$R_1 = 2.8125\text{ k}\Omega$$

4.24 SHUNT VOLTAGE REGULATOR

Design Specifications

(*i*) Input voltage = + 18 to + 22 V

(*ii*) Output voltage = + 12 to 15 V

(*iii*) Output current = 0 to 0.25 A.

Theory

Fig. 4.24 shows the schematic of a shunt voltage regulator circuit. To understand its operation, let us assume that the output voltage starts to increase, say, as a result of a decreased load current. When the load voltage starts to rise, the voltage across R_2 also increases. This is the feedback voltage for the regulator circuit and is essentially a sample of the output voltage. When the voltage across R_2 increases, *i.e.*, becomes more positive, the output of the op-amp becomes less positive because the voltage across R_2 is applied to the inverting input. This falling voltage on the output of the op-amp is the base voltage for Q, connected as an emitter follower, so the emitter voltage, and therefore the regulated output voltage, will decrease. Actually, the decrease merely offsets the original increase, so the output remains essentially constant. A similar closed-loop action would compensate for the decrease in output voltage and thereby maintaining it constant.

Another way to view the regulator action is to consider that the current in *Q* will increase in response to an increase in the regulated output voltage. This increased transistor current causes an increased voltage drop across *R*, thus returning the output voltage to its initial level. Because the current through *Q* increases and decreases to compensate for load voltage changes, the highest transistor current will occur during times when the load current is minimum.

Resistor *R* ultimately determines the maximum current that can be drawn from the regulator. If too much current is drawn, then *Q* is cut off and current is limited by *R*. Under these conditions, the output voltage is not regulated and will decrease with increasing load currents.

Circuit Diagram

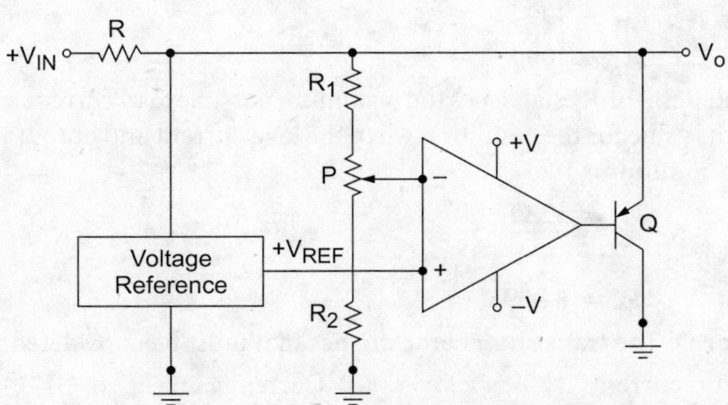

Fig. 4.24: Shunt voltage regulator circuit

Design Equations

The minimum and maximum voltage gains are

$$A_V(\text{min}) = \frac{V_O(\text{min})}{V_{REF}} \quad \text{and} \quad A_V(\text{max}) = \frac{V_O(\text{max})}{V_{REF}}$$

where V_{REF} is the reference voltage.

The minimum value for the potentiometer P is chosen much larger than the minimum equivalent load resistance R_L.

The resistor $R_1 = [A_V(\text{min}) - 1](R_2 + P)$ and $R_2 = \dfrac{PA_V(\text{min})}{A_V(\text{max}) - A_V(\text{min})}$

The resistor $R = \dfrac{V_{IN}(\text{min}) - V_O(\text{max})}{I_O(\text{max})}$

The power rating of R, $P_R = \dfrac{V_R^2}{R} = \dfrac{[V_{IN}(\text{max}) - V_O(\text{min})]^2}{R}$

The maximum collector current that the transistor will be expected to carry is

$$I_Q(\text{max}) = \frac{V_{IN}(\text{max}) - V_O(\text{min})}{R},$$

The required current gain for Q, $\beta = \dfrac{I_C}{I_B}$

The power dissipation of Q is $P_D = V_O(\text{max})I_Q(\text{max})$

Design Steps

Finding the error amplifier voltage gain: Since the design requires a variable output voltage, we will need to compute a range of error amplifier gains. We will assume a + 4 V reference voltage. The required voltage gains are,

$$A_V(\text{min}) = 3.0 \quad \text{and} \quad A_V(\text{max}) = 3.75$$

To adjust the gain of the error amplifier, we use a potentiometer P between R_1 and R_2 as shown in figure.

Selection of the potentiometer P: Choose $P(\text{min}) = 10R_L = 10\left(\dfrac{12 \text{ V}}{0.25 \text{ A}}\right)$

$$P(\text{min}) = 480\ \Omega$$

The maximum value can be few kΩ. Let us choose a 10 kΩ potentiometer for P.

Finding R_2: $R_2 = 40$ kΩ

Finding R_1: $R_1 = 100$ kΩ

Finding the value of R: Resistor R establishes the maximum possible load current from the shunt-regulator circuit. We can determine its value for the condition when the load current and output voltage are at maximum and the input voltage is at minimum. That is,

$$R = 12\ \Omega$$

The power rating of R is

$$P_R = 8.33 \text{ W}$$

Selection of transistor Q: The transistor characteristics that must be considered are

(*a*) Maximum collector current (*b*) Current gain (h_{FE} or β) (*c*) Breakdown voltages (*d*) Power dissipation

The maximum collector current that the transistor will be expected to carry can be computed as,

$$I_Q(\text{max}) = 833 \text{ mA}$$

In the circuit shown, the base current is provided by the output of the op-amp. The maximum current to be provided by the op-amp can be considered as one-fourth of the short-circuit output current rating of the op-amp, which, in the case of a 741, is listed in the data sheet as 20 mA. Therefore, we shall limit the output current and therefore the base current of Q to one-fourth of 20 mA, or 5 mA. The required current gain for *Q* can now be determined as

$$\beta = 167$$

The power dissipation of *Q* is

$$P_D = 12.5 \text{ W}$$

The collector-to-emitter breakdown voltage must be higher than V_O(max), *i.e.*, 15 V. Choose 2N6040 PNP transistor for the design, which is a Darlington power transistor that satisfies all the requirements.

4.25 LOW PASS FILTER

Design Specifications

(*i*) Cut-off frequency = 1 kHz (*ii*) $Q = 0.707$

(*iii*) Input impedance = 10 kΩ (min)

Theory

The low pass filter circuit is shown in Fig. 4.25. At low frequencies, the capacitors can be considered as open-circuit and the amplifier is connected as a simple voltage follower circuit. Resistor R_3 is included in the feedback loop to compensate for the effects of bias currents flowing through R_1 and R_2.

At high frequencies, the capacitive reactance is quite low and can be considered as short-circuit. The non-inverting input of the amplifier is essentially grounded. This should eliminate any chance of signals passing beyond this point. The junction of R_1 and R_2 is effectively connected to the output of the op-amp. Therefore, the low frequencies receive a voltage gain of about 1, and the high frequencies are severely attenuated, indicating a low pass filter action of the circuit.

Circuit Diagram

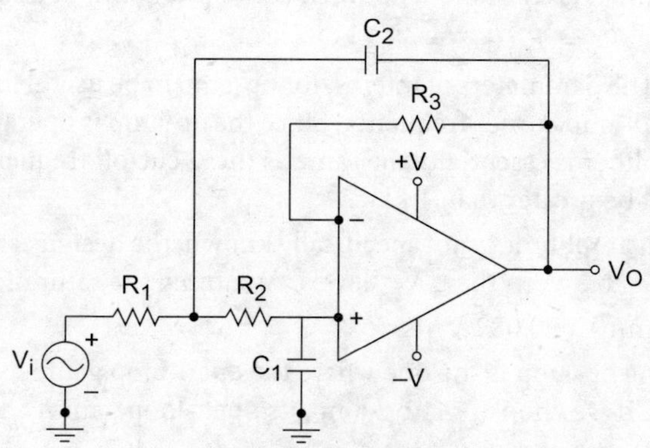

Fig. 4.25: Low pass filter circuit

Design Equations

The quality factor of the circuit is

$$Q = \frac{1}{2}\sqrt{\frac{C_2}{C_1}}$$

The ratio $\frac{C_2}{C_1} = 4Q^2$

The cut-off frequency

$$f_C = \frac{1}{2\pi R\sqrt{C_1 C_2}}$$

$$R = R_1 = R_2 = \frac{1}{2\pi f C_1\sqrt{C_1 C_2}}$$

The resistor R_1 must be larger than the minimum required input impedance.

$$R_3 = R_1 + R_2$$

Minimum required slew rate for the op-amp, slew rate(min) = $\pi f_C V_O$(max) where V_O(max) is the highest expected peak-to-peak output swing.

The minimum op-amp corner frequency, $f_{\text{break}}(\text{min}) = \frac{40 f_C}{A_{OL}}$

where A_{OL} is the low frequency open-loop gain of the op-amp.

Design Steps

Finding the ratio of $\frac{C_2}{C_1}$: The ratio of $\frac{C_2}{C_1}$ determines the Q of the circuit.

$$\frac{C_2}{C_1} = 2$$

Let $C_1 = 3300$ pF. Then $C_2 = 6600$ pF

Finding R_1 and R_2: Given, $f_C = 1$ kHz. Then

$$R = R_1 = R_2 = 34.1 \text{ k}\Omega$$

Finding the value of R_3: The DC resistances between each op-amp input and ground shall be equal.

$$R_3 = R_1 + R_2 = 68.2 \text{ k}\Omega$$

Selection of the op-amp: The parameters of interest for op-amp to be useful for low-pass filter application are bandwidth, slew rate, and op-amp corner frequency. Since the op-amp is operated as a voltage follower, the required bandwidth of the amplifier is essentially the same as the cut-off frequency, *i.e.*, $f_t(\text{min}) = f_C$. So, the bandwidth of the op-amp must be greater than 1 kHz.

Since the maximum output amplitude is not specifically known, the design can be done for worst case and assume that the signal will swing between the $\pm V_{\text{sat}}$ levels. Assuming the saturation limits as ± 13 V,

$$\text{Slew rate(min)} = 0.082 \text{ V}/\mu\text{s}$$

The corner frequency of the op-amp is the one where the open-loop gain has dropped to 70.7 percent of its low-frequency or DC value. If we choose 741 op-amp, its open-loop gain A_{OL} = 50,000. So, it must have a corner frequency greater than

$$f_{\text{break}} = 0.8 \text{ Hz}$$

Let us consider a 741 op-amp for this application. By referring to the graph of open-loop frequency response of 741, we can estimate the corner frequency as about 5Hz. Therefore 741 op-amp meet the requirements of the design.

4.26 HIGH PASS FILTER

Design Specifications

(*i*) Cutoff frequency = 200 Hz

(*ii*) $Q = 0.707$

(*iii*) Input impedance = 3 kΩ (min)

(*iv*) Highest input frequency = 5 kHz.

Theory

The schematic diagram of a high pass filter is shown in Fig. 4.26. The operation of the circuit can be understood by observing its behavior at very low and very high frequencies. At very low frequencies, the capacitors will have a high reactance and will begin to appear as open circuits. The amplifier acts as a unity gain circuit, but it has no input signal. At low frequencies, we expect little or no output signal.

At high frequencies, the capacitors will have a low reactance and will begin to appear as short circuits. Also, resistor R_2 has been removed, because it is connected between two points that have the same signal amplitude and phase, *i.e.*, input and output of a voltage follower. Because it has the same potential on both ends, it will have no current flow and is essentially open. The resulting equivalent circuit indicates that for high frequencies, the high-pass filter will act as a simple voltage follower.

Circuit Diagram

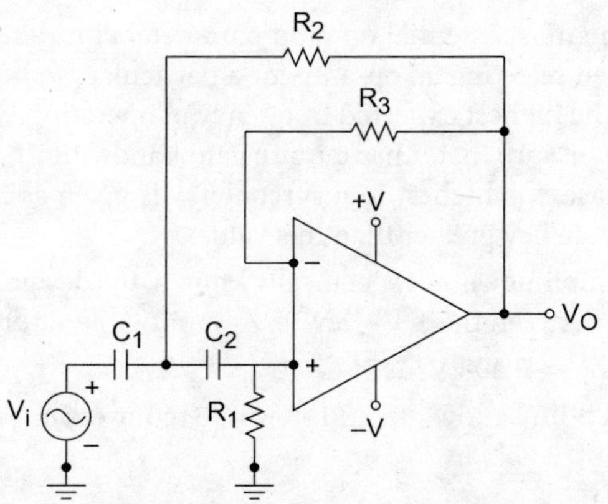

Fig. 4.26: High pass filter circuit

Design Equations

The quality factor of the circuit is

$$Q = \frac{1}{2}\sqrt{\frac{R_1}{R_2}}$$

The ratio $\dfrac{R_1}{R_2} = 4Q^2$

The resistor R_1 is larger than the minimum input impedance specified in the design criteria.

The cut-off frequency

$$f_C = \frac{1}{2\pi C\sqrt{R_1R_2}}$$

$$C = C_1 = C_2 = \frac{1}{2\pi f_C\sqrt{R_1R_2}}$$

Resistor R_3 is included to reduce the effects of the op-amp bias current that flows through R_1.

$$R_3 = R_1$$

Minimum required slew rate for the op-amp, slew rate(min) = $\pi f_C V_O$(max)
where V_O(max) is the highest expected peak-to-peak output swing.

Design Steps

Finding the ratio of $\frac{R_1}{R_2}$: The ratio of $\frac{R_1}{R_2}$ determines the Q of the circuit.

$$\frac{R_1}{R_2} = 2$$

Choosing R_2 as 10 kΩ, $R_1 = 20$ kΩ.

Finding the values of C_1 and C_2: Capacitors C_1 and C_2 are equal in value and can be computed as

$$C = C_1 = C_2 = 56.27 \text{ nF}$$

Finding the value of R_3: R_3 is set equal to R_1, or simply,

$$R_3 = 20 \text{ k}\Omega$$

Selection of the op-amp: The most essential op-amp parameters for designing a high-pass filter circuit are bandwidth and slew rate. When selecting an op-amp for a particular application, we must ensure that the amplifier gain is still adequate at the highest expected frequency of operation. Since the op-amp is configured for unity gain, we simply need to be sure that it has a unity gain bandwidth f_t that is higher than the highest input frequency. In the present case, the highest input frequency is given as 5 kHz. Therefore, the op-amps chosen must have a unity gain frequency greater than this value.

Since the maximum output amplitude is not specifically known, the design can be done for worst case and assume that the signal will swing between the ± V_{sat} levels. Assuming the saturation limits as ± 13 V,

$$\text{Slew rate(min)} = 0.408 \text{ V/}\mu\text{s}$$

Let us choose 741 op-amp as both bandwidth and slew rate requirements are within its capabilities.

4.27 BAND PASS FILTER

Design Specifications

(*i*) Resonant frequency f_r = 10 kHz
(*ii*) Q = 10
(*iii*) Voltage gain at f_r = 1
(*iv*) Bandwidth = 1 kHz.

Theory

The schematic diagram of a band pass filter is shown in Fig. 4.27. The operation of the circuit can be understood by observing its behavior at very low and very high frequencies. At very low frequencies, the capacitors will have a high reactance and will begin to appear as open circuits. Therefore, the low frequency components will not pass through the filter.

At high frequencies, the reactance of the capacitors will be low and the capacitors will act like short circuits. The high frequencies will be attenuated by the voltage divider action of R_1 and R_2. Additionally, the amplifier has zero resistance in the feedback loop, which causes voltage gain to be zero for the op-amp, *i.e.*, no output. At some intermediate frequency, determined by the component values, the gain of the amplifier will offset the loss in the voltage divider and the signals will be allowed to pass through.

Circuit Diagram

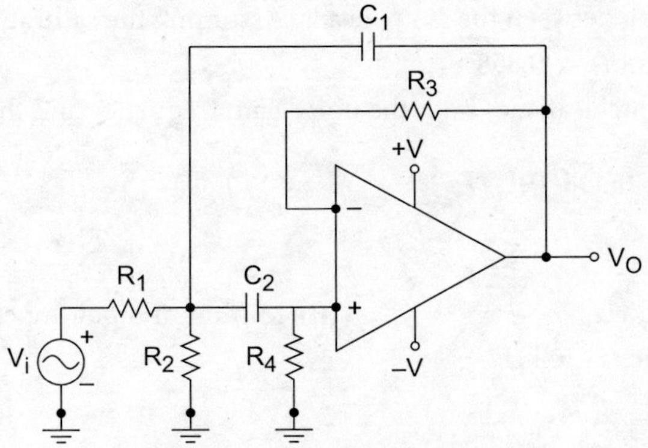

Fig. 4.27: Band pass filter circuit

Design Equations

The resistor $$R_1 = \frac{Q}{2\pi A_V f_r C}$$

where Q is the quality factor, f_r is the resonant frequency, A_V is the voltage gain and $C = C_1 = C_2$.

The resistor $$R_2 = \frac{R_1 A_V}{2(2\pi f_r C R_1 A_V)^2 - A_V}$$

$$R_3 = 2R_1$$

The resistor R_4 has no direct effect on the frequency response of the filter circuit. It is included to compensate for the effects of the op-amp bias current that flows through R_3. We try to keep equal the resistances between ground and the input pins of the op-amp. Therefore, R_4 is set equal to R_3.

The most essential op-amp parameters for designing a band-pass filter circuit are bandwidth and slew rate. The required bandwidth of the op-amp is determined by the highest frequency that must pass the circuit. This is the upper cutoff frequency f_H and can be approximated as

$$f_H = f_r + \frac{BW}{2}$$

The required bandwidth for the op-amp, $f_t = A_V f_H$

The minimum required slew rate for the op-amp, slew rate(min) = $\pi f_H V_O(\max)$

Design Steps

Finding C_1 and C_2: The choice of values for C_1 and C_2 will ultimately determine the values for the resistors. Choose $C_1 = C_2 = 1$ nF.

Finding the value of R_1: The resistor R_1 is given by

$$R_1 = 159.15\ \text{k}\Omega$$

Finding the value of R_2: The resistor $R_2 = 800\ \Omega$

Finding the value of R_3: The resistor $R_3 = 318.3\ \text{k}\Omega$

Finding the value of R_4: $R_4 = R_3 = 318.3\ \text{k}\Omega$

Selection of the op-amp: $f_H = 10.5\ \text{kHz}$

The required bandwidth for the op-amp is therefore

$$f_t = 10.5\ \text{kHz}$$

Since the maximum output amplitude is not specifically known, the design can be done for worst case and assume that the signal will swing between the $\pm V_{sat}$ levels. Assuming the saturation limits as $\pm$ 13 V,

$$\text{Slew rate(min)} = 0.858\ \text{V}/\mu\text{s}$$

Choose LM118 op-amp which satisfies both the unity-gain frequency and the slew rate requirements.

4.28 TWIN-T BAND REJECT FILTER

Design Specifications

(*i*) Center frequency = 5 kHz

(*ii*) Minimum input impedance = 10 kΩ

(*iii*) Highest input frequency = 20 kHz.

Theory

Fig. 4.28 shows an active filter that is based on the common twin "*T*" configuration. The twin-*T* gets its name from the two *RC T* networks in the input. Let us consider the lower ends of R_3 and C_3 to be grounded. This is a reasonable approximation, since the output impedance of an op-amp is generally quite low. The *T* circuit consisting of C_1, C_2, and R_3 is, by itself, a high pass filter. The second *T* network is made up of R_1, R_2 and C_3 and forms a low-pass filter. So, both low and high frequencies have a way to get to the non-inverting input of the op-amp and therefore to be passed through to the output. If, however, the cutoff frequencies of the two *T* networks do not overlap, there is a frequency f_r that results in a net voltage of 0 at the non-inverting terminal of the op-amp.

To understand this effect, we must also consider the phase shifts given to a signal as it passes through the two networks. At the center, or resonant, frequency f_r, the signal is shifted in the negative direction while passing through one *T* network. It receives the same amount of positive phase shift while passing through the other *T* network. These two shifted signals pass through equal impedances R_2 and X_{C2} to the non-inverting input. Thus, at any instant in time at the center frequency, the effective voltage on the non-inverting input is zero. The more the input frequency deviates from the center frequency, the less the cancellation effect. Thus, as we initially expected, this circuit rejects a band of frequencies and passes those frequencies that are higher or lower than the cutoff frequencies of the filter.

The op-amp offers a high impedance to the *T* networks, thus reducing the loading effects and therefore increasing the *Q* of the circuit. Additionally, by connecting the "ground" point of C_3 and R_3 to the output of the op-amp, the *Q* increases as a result of the feedback signal. At or very near the center frequency, almost zero signal makes it to the non-inverting input of the op-amp. Therefore, signal at the output of the op-amp is also very small. Under these conditions the output of the op-amp merely provides a ground, *i.e.*, low impedance return to ground for the *T* networks. For the other frequencies, the feedback essentially raises the impedance offered by C_3 and R_3 at a particular frequency. Therefore, they do not attenuate the off-resonance signals as much, which has the effect of narrowing the bandwidth or increasing the *Q*.

Resistor R_4 is to compensate for the voltage drops caused by the op-amp bias current flowing through R_1 and R_2. It is generally equal in value to the sum of R_1 and R_2.

Circuit Diagram

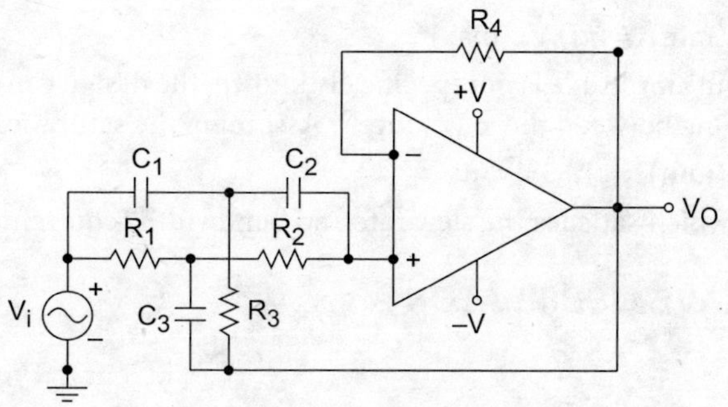

Fig. 4.28: Twin-T band reject filter circuit

Design Equations

The minimum value of resistor R_3, $R_3(\text{min}) = 2Z_{\text{in}}(\text{min})$, Z_{in} is the input impedance of the circuit. The upper limit for R_3 is established by the effects of op-amp bias currents and minimum practical values for C_1 to C_3. Although the bias currents do not actually flow through R_3, it is the value of R_3 that will determine the values for R_1 and R_2.

The value of C_1 is found as

$$C_1 = \frac{1}{4\pi f_r R_3}$$

where f_r is center or resonant frequency.

The capacitor C_2 is always the same value as C_1 and C_3 is twice that of C_1.

$$C_2 = C_1 \quad \text{and} \quad C_3 = 2C_1$$

Resistors R_1 and R_2 are equal in size and are twice the value of R_3.

$$R_1 = R_2 = 2R_3$$

Resistor R_4 helps to compensate for the voltage drop across R_1 and R_2 that is produced by the op-amp bias currents. To minimize the effects of the bias currents, R_4 is set equal to the sum of R_1 and R_2.

$$R_4 = R_1 + R_2$$

The required bandwidth for the op-amp, $f_t = A_V f_H$

where A_V is the voltage gain at f_r and is equal to 1 and f_H is the highest input frequency.

The minimum required slew rate for the op-amp, slew rate(min) = $\pi f_H V_O(\text{max})$

Design Steps

Finding the value of R_3: The value of R_3 is chosen based on the minimum input impedance requirement.

$$R_3(\text{min}) = 20\ \text{k}\Omega$$

Let us initially choose a value of 22 kΩ for R_3.

Finding the value for C_1: $C_1 = 723$ pF

Finding C_2 and C_3: $C_2 = 723$ pF and $C_3 = 1446$ pF

Finding R_1 and R_2: $R_1 = R_2 = 44$ kΩ

Finding R_4: $R_4 = 88$ kΩ

Selection of the op-amp: The most essential op-amp parameters for designing a band reject filter circuit are bandwidth and slew rate. The required bandwidth of the op-amp is determined by the highest frequency that must pass through the circuit.

$$f_t = A_V f_H = 1 \times 20 \text{ kHz} = 20 \text{ kHz}$$

The minimum required slew rate for the op-amp can be estimated as

$$\text{Slew rate(min)} = \pi f_H V_O(\text{max})$$

Since the maximum output amplitude is not specifically known, the design can be done for worst case and assume that the signal will swing between the $\pm V_{sat}$ levels. Assuming the saturation limits as ± 13 V,

$$\text{Slew rate(min)} = 1.633 \text{ V/}\mu\text{s}$$

Choose LM118 op-amp which satisfies the slew rate and bandwidth requirements.

4.29 TRIANGULAR WAVE OSCILLATOR

Design Specifications

(*i*) Frequency of oscillations = 1 kHz (*ii*) Ramp amplitude = ± 2 V.

Theory

Fig. 4.29 shows the schematic of an oscillator circuit that generates a dual ramp or triangle output. The heart of the circuit is amplifier A_1 which uses a capacitor as the feedback element. This operates as an integrator. Let us assume that the output of A_2 is at its $-V_{sat}$ level. Under these conditions, electrons will flow from the negative potential at the output of A_2 through R_1 and then C_1 as a charging current. The value of this current is determined by the voltage at the output of A_2 and the value of R_1. Since neither of these is changing at the moment, we assume the charging current is constant.

Whenever a capacitor is charged from a constant current source, the voltage across it accumulates linearly. Therefore, the voltage across C_1 will be increasing linearly, with the right side becoming more positive. Since the left side of C_1 is connected to a virtual ground point, the right side has a positive-going ramp with reference to ground. This is the output signal. When the positive-going ramp exceeds the upper threshold voltage of A_2, operating as a non-inverting voltage comparator, the output of A_2 will quickly switch to its $+V_{sat}$ level.

The electron flow through C_1 now reverses and flows from C_1 through R_1 toward the positive potential at the A_2 output. Again, the value of the current is constant and determined by R_1 and the voltage at the output of A_2. The voltage across C_1 will decay linearly until it passes through zero. It will then begin to charge at the same rate in the opposite polarity, producing the negative slope on the output. The output of A_1 continues to become more negative until it falls below the lower threshold voltage of A_2. At this time, the output of A_2 switches to its $-V_{sat}$ level and the cycle repeats.

Circuit Diagram

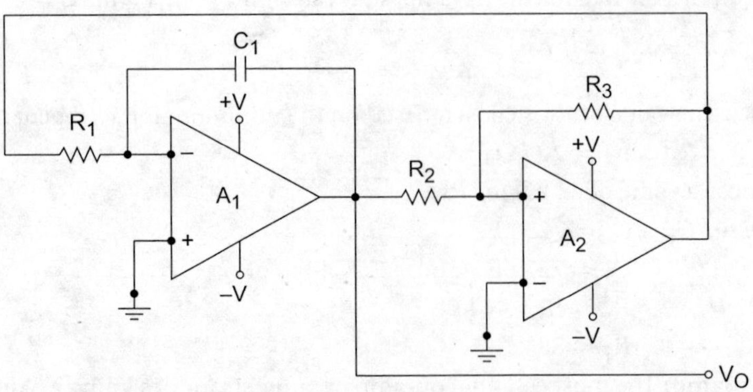

Fig. 4.29: Dual ramp or triangular wave oscillator circuit

Design Equations

Resistors R_2 and R_3 establish the threshold voltages for comparator A_2. These voltages in turn determine the output amplitude of the ramp voltage.

The threshold voltage, $\pm V_{th} = V_{sat}\dfrac{R_2}{R_3}$

The ratio of resistors, $\dfrac{R_2}{R_3} = \dfrac{V_{th}}{V_{sat}}$

The total period for one cycle, $T_1 + T_2 = \dfrac{1}{f_O}$

where f_O is the frequency of oscillations.

Since T_1 and T_2 are equal, $T_1 = T_2 = R_1C_1 \ln\left[\dfrac{V_{sat}+V_{th}}{V_{sat}-V_{th}}\right]$

$$R_1C_1 = \frac{T_{1,2}}{\ln[(V_{sat}+V_{th})/(V_{sat}-V_{th})]}$$

Assuming the rise and fall times are equal to 10% of one alternation of the triangular wave, slew rate (min) $= \dfrac{+V_{sat}-(-V_{sat})}{0.1T_{1,2}}$

Design Steps

Finding the values of R_2 and R_3:

$$\frac{R_2}{R_3} = \frac{V_{th}}{V_{sat}} = \frac{2}{13} = 0.154 \text{ V}$$

Choosing $R_2 = 47\text{ k}\Omega$

$R_3 = 305.2\text{ k}\Omega$

Finding R_1 and C_1: The total period for one cycle is

$$T_1 + T_2 = 1 \text{ ms}$$

Since T_1 and T_2 are equal, $T_1 = T_2 = 0.5$ ms

$$R_1C_1 = 1.612 \text{ ms}$$

Choosing $C_1 = 4.7$ nF, $R_1 = 343\text{ k}\Omega$

Selection of op-amp: The slew rate minimum of the op-amp = 0.52 V/μs

Choose LM118 op-amp which satisfies the slew rate requirements.

4.30 VOLTAGE CONTROLLED OSCILLATOR

Design Specifications

(*i*) Input voltage = 0 to 5 V DC

(*ii*) Ramp output voltage = ± 3 V (± 2 V minimum)

(*iii*) Frequency range = 0 to 4 kHz.

Theory

A voltage controlled oscillator (VCO) is an oscillator circuit whose frequency can be controlled or varied by a DC input voltage. This type of circuit is also called a voltage-to-frequency converter (VFC). The output waveform from the VCO may be sine, square, or other waveshape depending on the circuit design. Fig. 4.30 shows the schematic of a VCO circuit, which produces both triangle- and square-wave outputs. In both cases, the frequency is determined by the magnitude of the DC input voltage ($+ V_{IN}$).

The circuit's operation can be examined one stage at a time. The left most stage is basically an inverting summing amplifier with the feedback resistor replaced by a capacitor. The value of feedback current in an inverting amplifier is determined by the input voltage and the value of the input resistor(s). First, let us assume that diode D_4 is reverse-biased and acting as an open. Under these conditions, $+ V_{IN}$ and R_1 will determine the value of feedback current for A_1. Since V_{IN} is DC and R_1 does not change, the value of input current and, therefore, the feedback current will be constant. The feedback current must flow through C_1.

When a capacitor is charged with a constant current source, the resulting voltage will increase linearly. Because the charging current for C_1 is constant, we can expect a linear ramp of voltage across C_1. Since the left end of C_1 is connected to a virtual ground point, the other end connected to the output of the op-amp, will reflect the linear ramp voltage. The input voltage V_{IN} is positive, so we know that the output ramp will be increasing in the negative direction. A_2 is configured as a voltage comparator circuit with the upper and lower thresholds being established by diodes D_1 and D_2. As long as the ramp voltage is above the lower threshold point, established by D_1, the output of amplifier A_2 will remain at its negative limit ($- V_{sat}$).

Amplifier A_3 is connected as an inverting summing amplifier. One input comes from A_2 and receives a gain of – 2. The other input is provided by $+ V_{IN}$ and receives a gain of – 1. As long as the output of A_2 is at its $- V_{sat}$ level, diode D_3 will be forward-biased and this voltage will be coupled to the input of A_3. Clearly, this high-negative voltage will drive amplifier A_3 into saturation. That is, the output of A_3 will be at the $+ V_{sat}$ level regardless of the value of input voltage, $+ V_{IN}$. It is this $+ V_{sat}$ level on the output of A_3 that causes D_4 to remain in a reverse biased state. This circuit condition remains constant as long as the ramp voltage on the output of A_1 is above the lower threshold voltage of A_2.

Once the decreasing ramp voltage from A_1 falls below the lower threshold voltage of comparator A_1, the output of A_2 changes to its $+ V_{sat}$ level. This reverse biases diode D_3 and causes A_3 to act as a simple inverting amplifier with regard to the input voltage $+ V_{IN}$. A voltage level that is equal but opposite polarity to $+ V_{IN}$ is felt at the right end of R_4. Since R_4 is half as large as R_1 and has the same voltage applied, we can expect the current flow through R_4 to be twice as large as that through R_1 and in the opposite direction.

The current provided by R_4 not only cancels the input current provided via R_1 but supplies an equal but opposite current to C_1. That is, C_1 will now continue to charge at the same linear rate but in the opposite direction. The ramp voltage at the output of A_1 will rise linearly until it exceeds the upper threshold of A_2. Once the upper threshold has been exceeded, the output of A_2 switches to the $- V_{sat}$ level. This forces the output of A_3 to $+ V_{sat}$ and reverse-biases D_4. We are now back to the original circuit state, and the cycle repeats.

The frequency of operation is determined by the time it takes C_1 to charge to the threshold levels of A_2. Once the circuit components have been fixed, the only thing that determines frequency is the value of input voltage $+ V_{IN}$. This, of course, gives rise to the name voltage-controlled oscillator. A triangle-wave or double ramp signal may be taken from the output of A_1. The output of A_2 provides a square-wave output.

Circuit Diagram

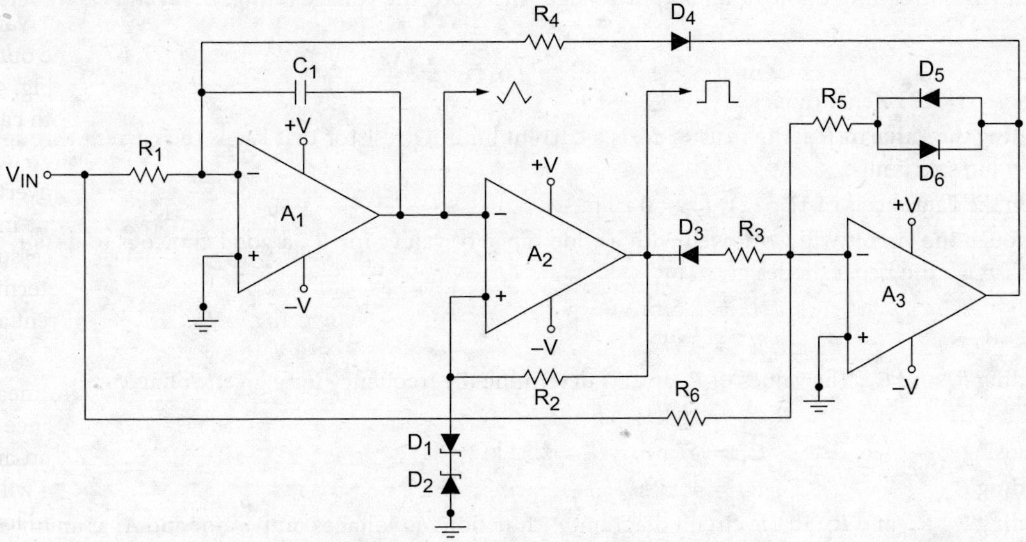

Fig. 4.30: Voltage controlled oscillator circuit

Design Equations

Voltage ratings of D_1 and D_2 are

$$V_{D1} = -V_{ro} + 0.6 \quad \text{and} \quad V_{D2} = +V_{ro} - 0.6$$

where V_{ro} is the amplitude of the ramp output voltage.

If I_{ZK} is the knee current of Zener, then the current through the Zener diodes, $I_Z = 10I_{ZK}$

The resistor $$R_2 = \frac{V_{sat} - V_Z - 0.6}{I_Z}$$

where V_{sat} is the highest expected saturation voltage for A_2, V_z is the lower of the two Zener voltages, if they are not equal.

The R_1C_1 time constant $= \dfrac{V_{IN}(\max)}{2f_{OH}V_{ro}}$

where $V_{IN}(\max)$ is the highest input voltage and f_{OH} is the highest frequency of oscillation.

The resistor R_4 must be exactly one-half of R_1, *i.e.*, $R_4 = \dfrac{R_1}{2}$

$$-\frac{R_5}{R_3} = -2 \quad \text{and} \quad -\frac{R_5}{R_6} = -1$$

Assuming the maximum switching time is 20% of time for one alternation,

$$\text{Slew rate (min)} = 40f_{OH}[+V_{sat} - (-V_{sat})]$$

Design Steps

Selection of the Zener diodes: The voltage rating of D_2 plus the forward voltage drop of D_1 determines the upper limit of the ramp output voltage. Similarly, the voltage rating of D_1 plus the forward voltage drop of D_2

determines the lower limit of the ramp output voltage. Therefore, the voltage ratings of D_1 and D_2 are selected as

$$V_{D1} = -2.4\text{ V} \quad \text{and} \quad V_{D2} = +2.4\text{ V}$$

Choose 1N5221 Zener diodes.

Finding the value of R_2: The resistor R_2 is a current limiting resistor that keeps the current through D_1 and D_2 within safe limits.

From the data sheets of 1N5221, $I_{ZK} = 0.25$ mA.

Although the circuit will work well with a wide range of values for R_2, a good choice is to design for a current through the Zener diodes given by

$$I_Z = 2.5\text{ mA}$$

$$R_2 = 4\text{ k}\Omega$$

Finding R_1 and C_1: The values of R_1 and C_1 determine the frequency for a given voltage.

$$R_1C_1 = 104.17\ \mu\text{s}$$

Choosing $C_1 = 47\text{ nF}, \quad R_1 = 2.22\text{ k}\Omega$

Finding R_4: $R_4 = 1.11\text{ k}\Omega$

Finding R_3, R_5, and R_6: In the circuit diagram, A_2 functions as voltage comparator and A_3 as an inverting summing amplifier. The amplifier A_3 receives two inputs, with one input provides a gain of – 2 and the other input has a gain of – 1.

$$-\frac{R_5}{R_3} = -2 \quad \text{and} \quad -\frac{R_5}{R_6} = -1$$

Choosing $R_5 = 10\text{ k}\Omega$, $R_3 = 5\text{ k}\Omega$ and $R_6 = 10\text{ k}\Omega$

Selection of diodes D_3 through D_6: Any general purpose diode such as 1N914A can be chosen for D_3 through D_6.

Selection of op-amps: The primary op-amp characteristic to be considered for this application is slew rate. The slew rate must be fast enough to allow the rise and fall times of A_2 and A_3 to be a small part of the time for one alternation.

$$\text{Slew rate (min)} = 4.16\text{ V}/\mu\text{s}$$

Choose LM118 op-amps which satisfies the slew rate requirements.

4.31 RC PHASE-SHIFT OSCILLATOR

Design Specification

Frequency of oscillations $f_o = 750$ Hz.

Theory

The RC phase shift oscillator using op-amp is shown in Fig. 4.31. It consists of an op-amp as an amplifier operating in inverting mode and 3 R-C sections cascaded as the feedback circuit. Since the op-amp is used in the inverting mode, any signal that appears at the inverting terminal is shifted by 180° at the output. An additional 180° phase shift required for oscillations is provided by the cascaded RC network. Thus, the total phase shift around the circuit is 360° or 0°. At some specific frequency, the phase shift of the cascaded RC network is exactly 180° and feedback factor is 1/29. If the gain of the amplifier $|A_V| = R_F/R_1 = 29$, the total loop gain of the circuit becomes 1. The circuit will oscillate at this specific frequency and is given by

$$f_o = \frac{1}{2\pi RC\sqrt{6}}$$

Circuit Diagram

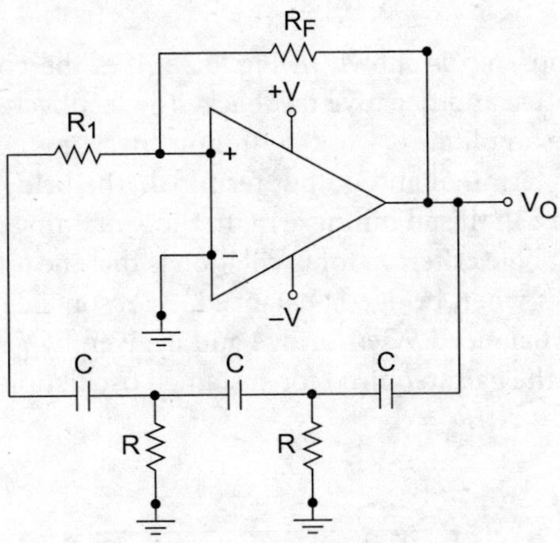

Fig. 4.31: RC phase-shift oscillator circuit

Design Equations

The frequency of oscillations, $f_o = \frac{1}{2\pi RC\sqrt{6}}$

The gain of inverting amplifier, $A_V = -\frac{R_F}{R_1}$

The unity-gain frequency f_t of op-amp shall be $10|A_V|f_O$

Design Steps

Finding R and C: Choosing $C = 0.1\ \mu F$, $R = 866\ \Omega$

Finding R_1: The resistor $R_1 = R = 866\ \Omega$

Finding R_F: The minimum gain required for the circuit is – 29.

$$-29 = -\frac{R_F}{866}$$

$$R_F = 25\ k\Omega$$

Use 33 kΩ potentiometer and adjust it in order to set the loop gain exactly 1.

Selection of op-amp: The GBW or unity-gain frequency of op-amp shall be at least 10 times larger than the GBW of the circuit.

$$f_t = 217.5\ kHz$$

We can choose 741 op-amp which has a unity-gain frequency of about 1 MHz and slew rate of 0.5 V/μs.

4.32 WIEN-BRIDGE OSCILLATOR

Design Specification

Frequency of oscillations $f_o = 500$ Hz.

Theory

The Wien-bridge oscillator using op-amp is shown in Fig. 4.32. It is the commonly used audio frequency oscillator which employs both positive and negative feedback. The feedback signal is connected to the non-inverting input terminal so that the amplifier is working in non-inverting mode. The Wien-bridge circuit is connected between amplifier input terminal and output terminal. The bridge has a series R_1-C_1 network in one arm and a parallel R_2-C_2 network in the adjoining arm. In the remaining two arms of the bridge, resistors R_3 and R_4 are connected. The phase angle criterion for oscillation is that the total phase shift around the circuit must be zero. This condition occurs when bridge is balanced. At resonance, the frequency of oscillation is exactly the resonance frequency of balanced Wien-bridge and is given by $f_o = 1/(2\pi RC)$, where $R = R_1 = R_2$ and $C = C_1 = C_2$. At this frequency, the gain required for sustained oscillations is 3. It is provided by the non-inverting amplifier with gain $A_V = 1 + R_3/R_4 = 3$.

Circuit Diagram

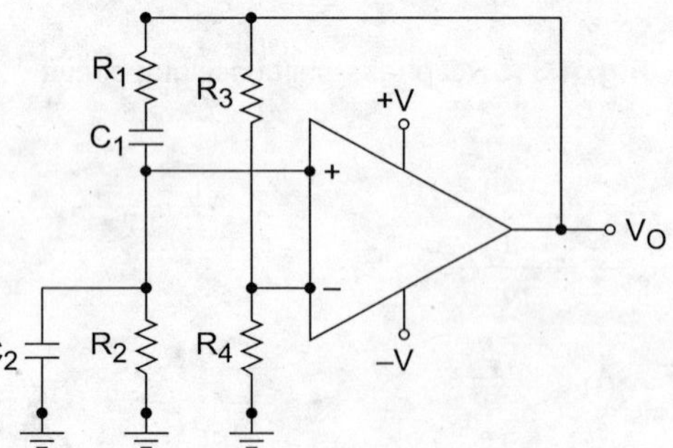

Fig. 4.32: Wien-bridge oscillator circuit

Design Equations

The frequency of oscillations, $f_o = \dfrac{1}{2\pi RC}$, $R = R_1 = R_2$ and $C = C_1 = C_2$

The gain of non-inverting amplifier, $A_V = 1 + \dfrac{R_3}{R_4}$

The unity-gain frequency f_t of op-amp shall be $10|A_V|f_O$

Design Steps

Finding R_1, R_2, C_1 and C_2: Choosing $C = C_1 = C_2 = 0.01$ μF, $R = R_1 = R_2 = 31.83$ kΩ

Finding R_3 and R_4: The gain required for the circuit is 3.

$$A_V = 1 + \frac{R_3}{R_4}$$

$$3 = 1 + \frac{R_3}{R_4} \quad \text{or} \quad \frac{R_3}{R_4} = 2$$

Choosing $R_4 = 10\ \text{k}\Omega,\quad R_3 = 20\ \text{k}\Omega$

Selection of op-amp: The GBW or unity-gain frequency of op-amp shall be at least 10 times larger than the GBW of the circuit.

$$f_t = 15\ \text{kHz}$$

We can choose 741 op-amp which has a unity-gain frequency of about 1 MHz and slew rate of 0.5 V/μs.

4.33 HARTLEY OSCILLATOR

Design Specification

Frequency of oscillations f_o = 400 kHz.

Theory

Hartley oscillator circuit using op-amp is shown in Fig. 4.33. It generally facilitates the gain adjustment using the feedback resistance and internal resistance. The gain can be expressed by the equation $A_V = -R_F/R$ and the circuit provides good frequency stability. The main advantage of using op-amp is that the gain of the oscillator can be individually adjusted using the feedback resistor R_F and input resistor R. The frequency of oscillations is given as

$$f_o = \frac{1}{2\pi(L_{eq}C)^{1/2}}, \quad L_{eq} = L_1 + L_2 + 2M$$

where M is the mutual inductance between L_1 and L_2.

The feedback factor $\beta = L_2/L_1$ and therefore, to generate oscillations from this circuit, amplifier gain must be selected greater than or, at least, equal to the ratio of two inductances, *i.e.*,

$$A_v = \frac{L_1}{L_2}$$

Considering the mutual inductance, the gain becomes

$$A_v = \frac{L_1 + M}{L_2 + M}$$

Circuit Diagram:

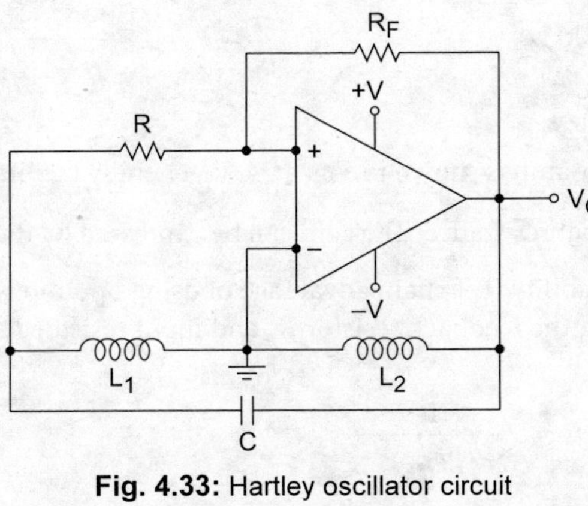

Fig. 4.33: Hartley oscillator circuit

Design Equations

The frequency of oscillations, $f_o = \frac{1}{2\pi(L_{eq}C)^{1/2}}$, $L_{eq} = L_1 + L_2$, neglecting mutual inductance M.

$$L_{eq} = \frac{1}{(2\pi f_o)^2 C}$$

The feedback factor $\beta = \frac{L_2}{L_1}$

The minimum gain required, $|A_V| = \frac{1}{\beta} = \frac{R_F}{R}$

The unity-gain frequency f_t of op-amp shall be $10|A_V| \cdot f_O$

Design Steps

Finding L_1, L_2, and C: Choosing $C = 0.01$ µF, $L_{eq} = 15.8$ µH

Let $L_1 = 12$ µH and so $L_2 = 3.8$ µH

Finding R_F and R: The feedback factor $\beta = 0.317$

Therefore the minimum gain required, $|A_V| = 3.16$

$$|A_V| = \frac{R_F}{R} = 3.16$$

Choose $R_F = 50$ kΩ. Then $R = 15.82$ kΩ

Selection of op-amp: The GBW or unity-gain frequency of op-amp shall be at least 10 times larger than the GBW of the circuit.

$$f_t = 12.64 \text{ MHz}$$

We can choose LM318 op-amp which has a unity-gain frequency of 15 MHz and slew rate of 70 V/µs.

4.34 COLPITTS OSCILLATOR

Design Specification

Frequency of oscillations $f_o = 200$ kHz.

Theory

Colpitts oscillator circuit using op-amp is shown in Fig. 4.34. It generally facilitates the gain adjustment using the feedback resistance and internal resistance. The gain can be expressed by the equation $A_V = -\frac{R_F}{R}$ and the circuit provide good frequency stability. The main advantage of using op-amp is that the gain of the oscillator can be individually adjusted using the feedback resistor R_F and input resistor R. The frequency of oscillations is given as

$$f_o = \frac{1}{2\pi(LC_{eq})^{1/2}}$$

$$C_{eq} = \frac{C_1 C_2}{C_1 + C_2}$$

The feedback factor $\beta = \frac{C_2}{C_1}$ and therefore, to generate oscillations from this circuit, amplifier gain must be selected greater than or, at least, equal to the ratio of two capacitors, *i.e.*,

$$A_v = \frac{C_1}{C_2}$$

Circuit Diagram

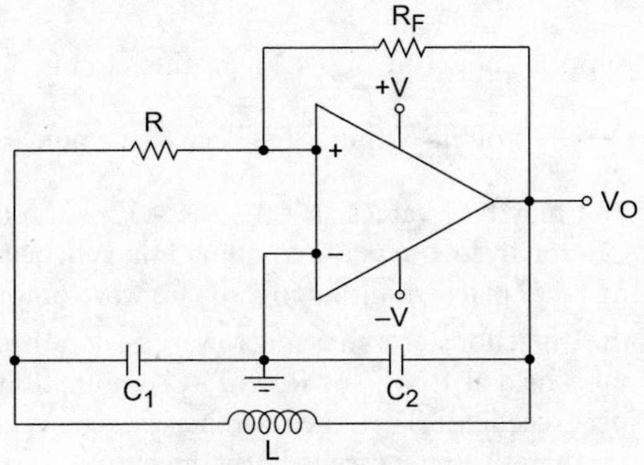

Fig. 4.34: Colpitts oscillator circuit

Design Equations

The frequency of oscillations, $f_o = \dfrac{1}{2\pi(LC_{eq})^{1/2}}, \quad C_{eq} = \dfrac{C_1 \cdot C_2}{(C_1 + C_2)}$

$$C_{eq} = \frac{1}{(2\pi f_o)^2 L}$$

The feedback factor $\beta = \dfrac{C_2}{C_1}$

The minimum gain required, $|A_V| = \dfrac{1}{\beta} = \dfrac{R_F}{R}$

The unity-gain frequency f_t of op-amp shall be $10|A_V| \cdot f_O$

Design Steps

Finding C_1, C_2, and L: Choosing $L = 10\ \mu H$, $C_{eq} = 63$ nF

Let $C_1 = 180$ nF and so $C_2 = 96.92$ nF

Finding R_F and R: The feedback factor $\beta = 0.538$

Therefore, the minimum gain required, $|A_V| = 1.86$

$$|A_V| = \frac{R_F}{R} = 1.86$$

Choose $R_F = 50\ k\Omega$. Then $R = 27\ k\Omega$

Selection of op-amp: The GBW or unity-gain frequency of op-amp shall be at least 10 times larger than the GBW of the circuit.

$$f_t = 3.72 \text{ MHz}$$

We can choose LM318 op-amp which has a unity-gain frequency of 15MHz and slew rate of 70 V/μs.

4.35 RELAXATION OSCILLATOR

Design Specification

Frequency of oscillations $f_o = 1$ kHz.

Theory

A relaxation oscillator using op-amp is shown in Fig. 4.35. In this circuit the op-amp is operated in the saturation region. A fraction $\beta = \frac{R_2}{R_1 + R_2}$ of the output is fed back to the non-inverting input terminal. Thus, the reference voltage V_{ref} is βV_o and may take values as $+ V_{sat}$ or $- V_{sat}$. The output is also fed back to the inverting input terminal after integration by low-pass RC combination. Whenever input at the inverting input terminal just exceeds V_{ref} switching takes place, resulting in a square wave output.

When the output is at $+ V_{sat}$, the capacitor starts charging towards $+ V_{sat}$ through resistance R. The voltage at the non-inverting input terminal is held at $+ \beta V_{sat}$ by R_1 and R_2 combination. This condition continues as the charge on C rises, until it has just exceeded $+ \beta V_{sat}$, the reference voltage. When the voltage at the inverting input terminal becomes just greater than this reference voltage, the output is driven to $- V_{sat}$. At this instant, the voltage on the capacitor is $+ \beta V_{sat}$. It begins to discharge through R. When the output voltage switches to $- V_{sat}$, the capacitor charges more and more negatively until its voltage just exceeds $- \beta V_{sat}$. The output switches back to $+ V_{sat}$. The cycle repeats itself as shown to generate square wave output with time period $T = T_C + T_D = T_{on} + T_{off} = 2RC \ln\left[\frac{1+\beta}{1-\beta}\right]$.

Circuit Diagram and Voltage Waveforms

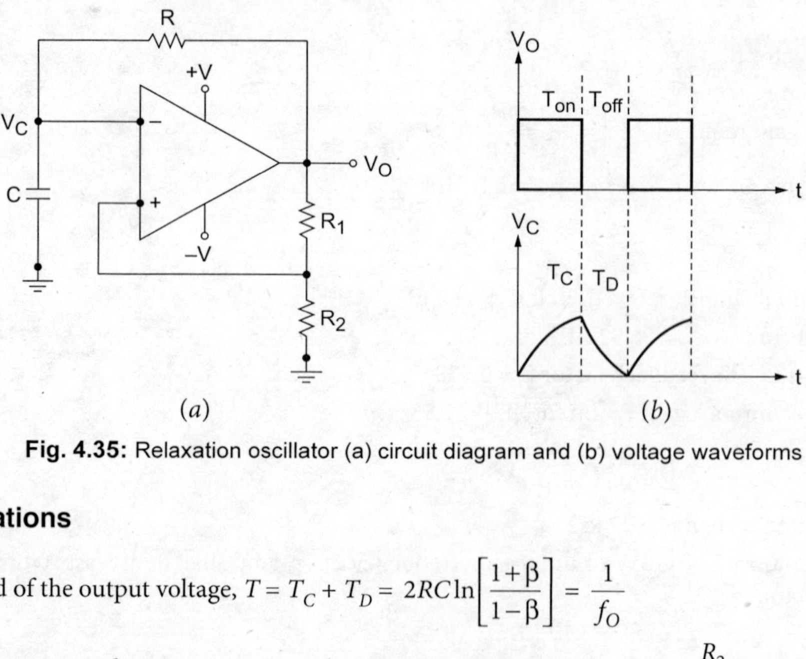

Fig. 4.35: Relaxation oscillator (a) circuit diagram and (b) voltage waveforms

Design Equations

The time period of the output voltage, $T = T_C + T_D = 2RC \ln\left[\frac{1+\beta}{1-\beta}\right] = \frac{1}{f_O}$

where T_C is the capacitor charging time, T_D is the capacitor discharge time, $\beta = \frac{R_2}{R_1 + R_2}$ is the feedback factor and f_O is the frequency of oscillations.

The slew rate minimum of op-amp $= \pi f_O V_O(\text{max})$

where $V_O(\text{max})$ is the maximum output signal amplitude.

Design Steps

$$T = \frac{1}{10^3} = 1 \text{ ms}$$

For $R_1 = R_2$, $\quad \beta = 0.5$

and $\quad T = 2.2\,RC$

Choose $C = 0.1\ \mu\text{F}$. Then $R = 4.55\ \text{k}\Omega$

Let $\quad R_1 = R_2 = 10\ \text{k}\Omega$

Choose 741 op-amp with saturation voltages of $V_{sat} = \pm 13$ V

$$\text{Slew rate (min.)} = 0.082\ \text{V}/\mu\text{s}$$

This value is within the rated value 0.5 V/μs of 741 op-amp. Hence 741 op-amp is suitable for the present design.

4.36 SCHMITT TRIGGER

Design Specifications

(*i*) UTP = + 3 V (*ii*) LTP = – 3 V

(*iii*) Input - sinusoidal signal of 5 V peak amplitude and 1 kHz frequency.

Theory

The Schmitt trigger or regenerative comparator using op-amp is shown in Fig. 4.36(a). The input voltage is applied to the inverting terminal and the feedback voltage to the non-inverting terminal of the op-amp. The input voltage V_i triggers the output V_o every time it exceeds the voltage levels, upper threshold voltage (V_{UTP}) and lower threshold voltage (V_{LTP}). The difference between these two is defined as the hysteresis, V_H, *i.e.*, $V_H = V_{UTP} - V_{LTP}$.

For $V_o = +V_{sat}$, the voltage at non-inverting input terminal is

$$V_{ref} \cdot \frac{R_2}{R_1 + R_2} + V_{sat} \cdot \frac{R_1}{R_1 + R_2} = V_{UTP}$$

As long as V_i is less than V_{UTP}, the output V_o remains constant at $+V_{sat}$. When V_i is just greater than V_{UTP}, the output regeneratively switches to $-V_{sat}$ and remains at this level as long as $V_i > V_{UTP}$ as shown in waveforms. For $V_o = -V_{sat}$, the voltage at the non-inverting input terminal is

$$V_{ref} \cdot \frac{R_2}{R_1 + R_2} - V_{sat} \cdot \frac{R_1}{R_1 + R_2} = V_{LTP}$$

The input voltage V_i must be less than V_{LTP} in order to cause V_o to switch from $-V_{sat}$ to $+V_{sat}$. A regenerative transition takes place as shown in Fig. 4.36(*b*) and the output V_o returns from $-V_{sat}$ to $+V_{sat}$ almost instantaneously. The transfer/hysteresis characteristics are shown in Fig. 4.36(*c*). The hysteresis V_H is given as

$$V_H = V_{UTP} - V_{LTP} = 2V_{sat} \cdot \frac{R_1}{R_1 + R_2}$$

Circuit Diagram and Characteristics

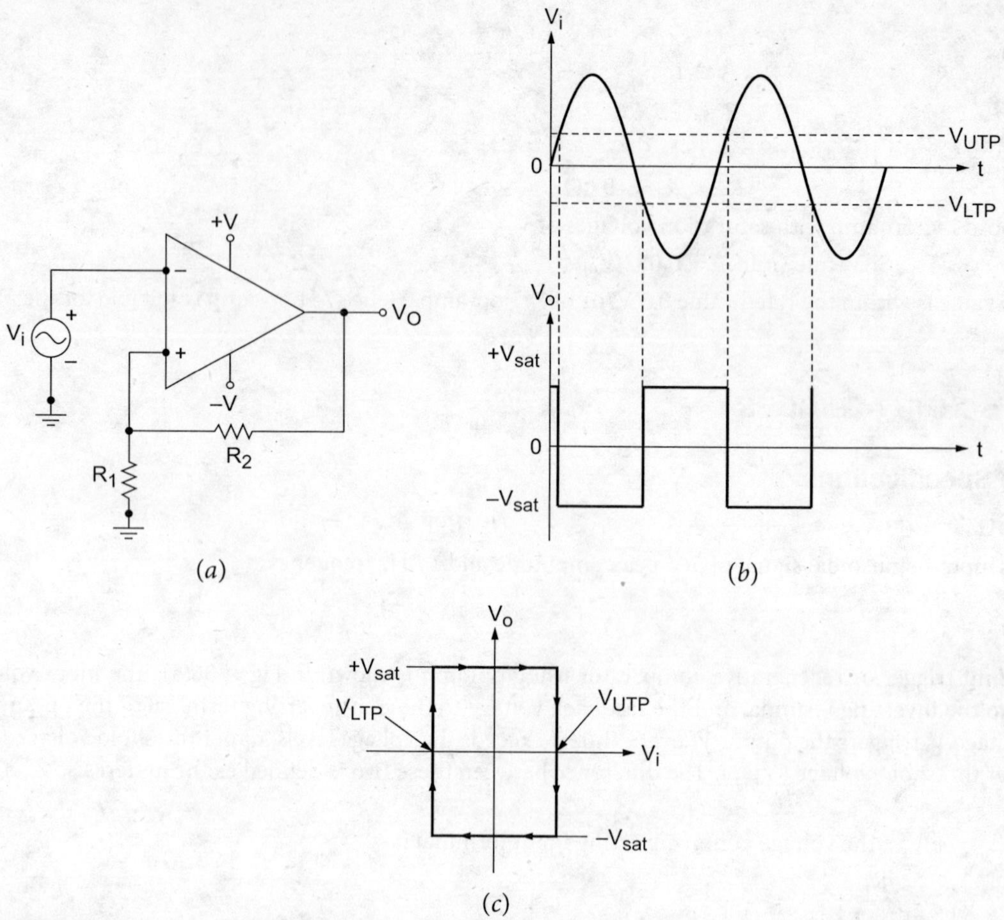

Fig. 4.36: Schmitt trigger (a) circuit diagram, (b) input-output characteristics, and (c) hysteresis characteristics

Design Equations

The upper trigger point (UTP) is given by

$$V_{\text{UTP}} = V_{\text{ref}} \cdot \frac{R_2}{R_1 + R_2} + V_{\text{sat}} \cdot \frac{R_1}{R_1 + R_2}$$

where V_{ref} is the reference voltage and V_{sat} is the saturation voltage of op-amp. For $V_{\text{ref}} = 0$,

$$V_{\text{UTP}} = +V_{\text{sat}} \cdot \frac{R_1}{R_1 + R_2}$$

The lower trigger point (LTP) is given by

$$V_{\text{LTP}} = V_{\text{ref}} \cdot \frac{R_2}{R_1 + R_2} - V_{\text{sat}} \cdot \frac{R_1}{R_1 + R_2}$$

For $V_{\text{ref}} = 0$,

$$V_{\text{LTP}} = -V_{\text{sat}} \cdot \frac{R_1}{R_1 + R_2}$$

The slew rate minimum of op-amp = $\pi f V_O(\text{max})$
where f is the frequency of the output signal and $V_O(\text{max})$ is the maximum output signal amplitude.

Design Steps

Assuming 741 op-amp, the saturation voltages for supply voltages of ± 15 V are V_{sat} = ± 13 V.

We have

$$V_{UTP} + V_{LTP} = 2V_{ref} \cdot \frac{R_2}{R_1 + R_2}$$

$$+3 + (-3) = 0$$

So,

$$V_{ref} = 0$$

$$V_{UTP} - V_{LTP} = 2V_{sat} \cdot \frac{R_1}{R_1 + R_2}$$

$$+3 - (-3) = 6 = \frac{26R_1}{R_1 + R_2}$$

Choose R_1 = 1 kΩ. Then R_2 = 3.3 kΩ

$$\text{Slew rate (min.)} = 0.082 \text{ V/}\mu\text{s}$$

This value is within the rated value 0.5 V/µs of 741 op-amp. Hence 741 op-amp is suitable for the present design.

4.37 ASTABLE MULTIVIBRATOR

Design Specifications

(*i*) Frequency of oscillations f_o = 1 kHz (*ii*) Duty cycle = 40%.

Theory

An astable multivibrator using op-amp is shown in Fig. 4.37. In this circuit the op-amp is operated in the saturation region. A fraction $\beta = \dfrac{R_2}{R_1 + R_2}$ of the output is fed back to the non-inverting input terminal. Thus, the reference voltage V_{ref} is βV_o and may take values as $+V_{sat}$ or $-V_{sat}$. The output is also fed back to the inverting input terminal after integration by low-pass $R_F C$ combination. Whenever input at the inverting input terminal just exceeds V_{ref}, switching takes place, resulting in a square wave output. In an astable multivibrator, both the states are quasi-stable.

When the output is at $+V_{sat}$, the diode D_1 is forward biased and capacitor starts charging towards $+V_{sat}$ through resistance R_{F1}. The voltage at the non-inverting input terminal is held at $+\beta V_{sat}$ by R_1 and R_2 combination. This condition continues as the charge on C rises, until it has just exceeded $+\beta V_{sat}$, the reference voltage. When the voltage at the inverting input terminal becomes just greater than this reference voltage, the output is driven to $-V_{sat}$ and diode D_2 is forward biased. At this instant, the voltage on the capacitor is $+\beta V_{sat}$. It begins to discharge through R_{F2}. When the output voltage switches to $-V_{sat}$, the capacitor charges more and more negatively until its voltage just exceeds $-\beta V_{sat}$. The output switches back to $+V_{sat}$. The cycle repeats itself as shown to generate square wave output with time period $T = T_{on} + T_{off} = 1.1C(R_{F1} + R_{F2})$. In asymmetrical astable multivibrators, the charging and discharging time of capacitor is made unequal to get asymmetrical square wave with different T_{on} and T_{off}.

Circuit Diagram and Voltage Waveforms

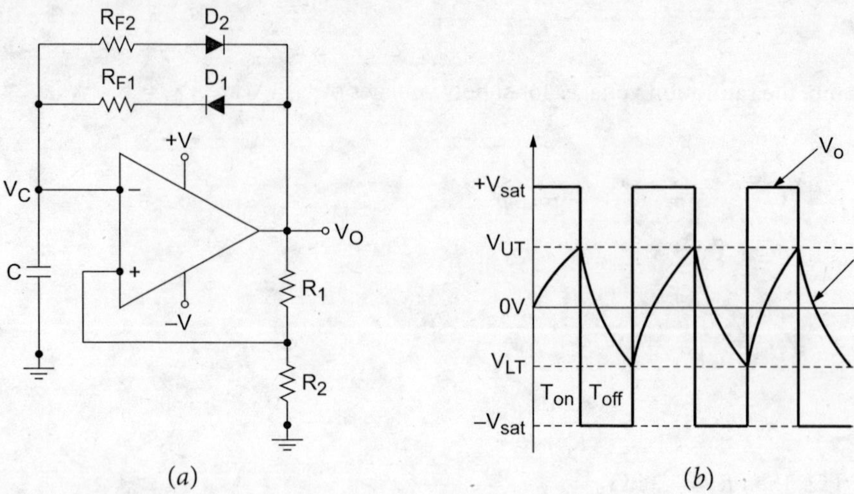

Fig. 4.37: Astable multivibrator (a) circuit diagram and (b) capacitor and output voltage waveforms

Design Equations

The feedback factor, $\beta = \dfrac{R_2}{R_1 + R_2}$

Time period of the symmetrical output square wave, $T = T_{on} + T_{off} = 2RC\ln\left[\dfrac{1+\beta}{1-\beta}\right]$

where T_{on} and T_{off} are the ON and OFF periods during which capacitor charging and discharging occurs and the output will be high and low, respectively.

$$R = R_{F1} \quad \text{or} \quad R_{F2}$$

For asymmetrical astable multivibrator,

$$T_{on} = 1.1R_{F1}C \quad \text{and} \quad T_{off} = 1.1R_{F2}C$$

$$T = \frac{1}{f_O}, f_O \text{ is the frequency of oscillations.}$$

Duty cycle, $D = \dfrac{T_{on}}{T_{on} + T_{off}} = \dfrac{T_{on}}{T}$

V_{UT} and V_{LT} are the upper and lower threshold voltages and V_{sat} is the op-amp saturation voltage.

The slew rate minimum of op-amp $= \pi f_O V_O(\text{max})$

where $V_O(\text{max})$ is the maximum output signal amplitude.

Design Steps

From the given design specifications, the astable multivibrator to be designed is of asymmetrical type.

$$T = \frac{1}{f_O} = \frac{1}{10^3} = 1 \text{ ms}$$

$$D = 40\% = 0.4$$

$$T_{on} + T_{off} = 10^{-3} \quad \text{and} \quad \frac{T_{on}}{T_{on} + T_{off}} = 0.4$$

$$T_{on} = 0.4 \text{ ms} \quad \text{and} \quad \text{hence } T_{off} = 0.6 \text{ ms}$$

Let $R_1 = R_2 = 10\ \text{k}\Omega$. So, $\beta = \frac{R_2}{R_1 + R_2} = 0.5$ and $T = 2.2RC$

From the circuit diagram, $T_{on} = 0.4 \text{ ms} = 1.1R_{F1}C$ and $T_{off} = 0.6 \text{ ms} = 1.1R_{F2}C$

Let $C = 0.1\ \mu\text{F}$.

Then, $R_{F1} = 3.64\ \text{k}\Omega$

$R_{F2} = 5.45\ \text{k}\Omega$

Choose 1N4001 diodes for D_1 and D_2.

Choose 741 op-amp with saturation voltages of $V_{sat} = \pm 13$ V.

Slew rate (min.) = 0.082 V/μs

This value is within the rated value 0.5 V/μs of 741 op-amp. Hence 741 op-amp is suitable for the present design.

4.38 MONOSTABLE MULTIVIBRATOR

Design Specification

Frequency of oscillations $f_o = 1$ kHz

Theory

The monostable multivibrator, also called as one-shot multivibrator, is shown in Fig. 4.38. It always has one stable state. In response to an external trigger pulse, the circuit produces a single pulse of specified duration. When an external trigger is applied, the output state changes and the new state is called quasi-stable state. The circuit remains in this state for a fixed interval of time and then it returns to the original state after this interval. This time interval is determined by the discharging of the capacitor. The time period of quasi-stable state is given by,

$$T = 0.69RC_1$$

The diode D_1 connected across the capacitor C_1 is called the clamping diode. It clamps the capacitor voltage to 0.7 V when the output is at $+V_{sat}$. A negative triggering pulse is applied to the non-inverting terminal of op-amp through R_3-C_2 differentiator circuit and diode D_2. To understand the operation of the circuit, assume that the output V_O is at $+V_{sat}$, *i.e.*, in its stable state. The diode D_1 conducts and the voltage across the capacitor V_C gets clamped to 0.7 Volts. The voltage at the non-inverting terminal is controlled by the voltage divider circuit consisting of R_1 and R_2. The voltage at the non-inverting terminal of the op-amp is

$$V_+ = \beta V_O$$

where $$\beta = \frac{R_2}{R_1 + R_2}$$

If a negative trigger of amplitude V_T is applied to the non-inverting terminal, so that the effective voltage at this terminal is less than 0.7 V, then the output of the op-amp changes its state from $+V_{sat}$ to $-V_{sat}$. The diode is now reverse biased and the capacitor C_1 starts charging exponentially to $-V_{sat}$ through resistance R. The voltage at the non-inverting terminal is now $-\beta V_{sat}$. When the capacitor voltage becomes just slightly more negative than $-\beta V_{sat}$, the output of the op-amp changes its state back to $+V_{sat}$. The capacitor now starts charging towards $+V_{sat}$ through R until V_C reaches 0.7 V as capacitor gets clamped to the voltage.

The time period of the output voltage waveform, $T = \dfrac{1}{f_O} \approx RC_1 \ln\left[\dfrac{1}{1-\beta}\right]$

where f_O is the frequency of oscillations and β is the feedback factor $= \dfrac{R_2}{R_1 + R_2}$

For $R_1 = R_2$, $\beta = 0.5$ and $T = 0.69RC_1$

The trigger pulse width, $t_p = R_3C_2$

Circuit Diagram and Waveforms

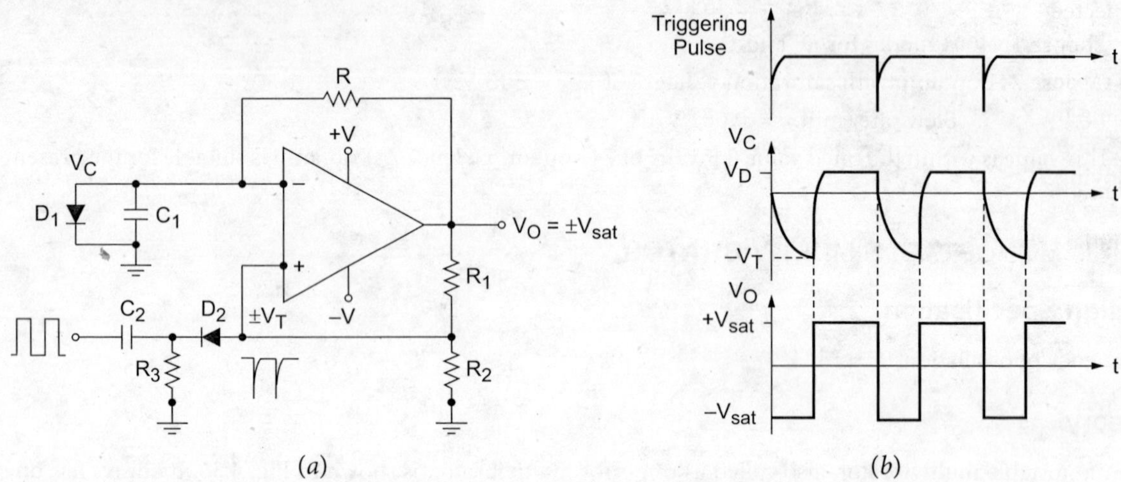

Fig. 4.38: Monostable multivibrator (a) circuit diagram and (b) waveforms

Design Equations

The time period of the output voltage waveform, $T = \dfrac{1}{f_O} \approx RC_1 \ln\left[\dfrac{1}{1-\beta}\right]$

where f_O is the frequency of oscillations and β is the feedback factor $= \dfrac{R_2}{R_1 + R_2}$

For $R_1 = R_2$, $\beta = 0.5$ and $T = 0.69RC_1$

The trigger pulse width, $t_p = R_3C_2$

Design Steps

Let $R_1 = R_2 = 10\ \text{k}\Omega$

Given, $f_O = 1$ kHz and $T = 1\ \text{ms} = 0.69RC_1$

Let $C_1 = 0.1\ \mu\text{F}$. Then $R = 14.5\ \text{k}\Omega$

The trigger pulse width, $t_p << T$

Let $t_p = \dfrac{T}{100} = 0.01$ ms

$$R_3C_2 = 0.01 \text{ ms}$$

Choose $C_2 = 0.01\ \mu\text{F}$. Then $R_3 = 1\ \text{k}\Omega$

Choose 1N4001 diodes for D_1 and D_2.

4.39 ASYMMETRICAL ASTABLE MULTIVIBRATOR USING 555

Design Specifications

(*i*) Frequency of oscillations = 1 kHz (*ii*) Duty cycle = 60%.

Theory

Figure 4.39(*a*) shows the astable or free-running multivibrator using 555. It has no stable states and continually switches between two quasi-stable states without the application of any trigger. The internal block diagram and various waveforms are given in Figs. 4.39(*b*) and 4.39(*c*) respectively. The trigger and threshold pins are connected together and hence there is no need for an external trigger pulse.

Assume initially that the 555 output is in the high state. At this point, the discharge transistor Q_1 is off and capacitor C charges toward V_{CC} through R_A and R_B. Eventually, the capacitor voltage will exceed $2/3V_{CC}$ causing the threshold comparator to trigger the flip-flop. This will turn on the discharge transistor and make the 555 output go low. The discharge transistor effectively places the upper end of R_B at ground potential, removing R_A and V_{CC} from consideration. C now discharges through R_B toward 0. Eventually, the capacitor voltage will drop below $1/3V_{CC}$. This will fire the trigger comparator, which in turn places the circuit back to its initial state, and the cycle will repeat.

The frequency of oscillation depends only on R_A, R_B and C. The ON and OFF time periods of output square wave are

$$T_{ON} = 0.69(R_A + R_B)C$$
$$T_{OFF} = 0.69\, R_B C$$

This results in a frequency of

$$f = \frac{1.38}{(R_A + 2R_B)C}$$

The duty cycle is normally defined as the ratio of T_{ON} to T.

$$\text{Duty cycle} = \frac{R_A + R_B}{R_A + 2R_B}$$

Circuit Diagram, Internal Block Diagram and Waveforms

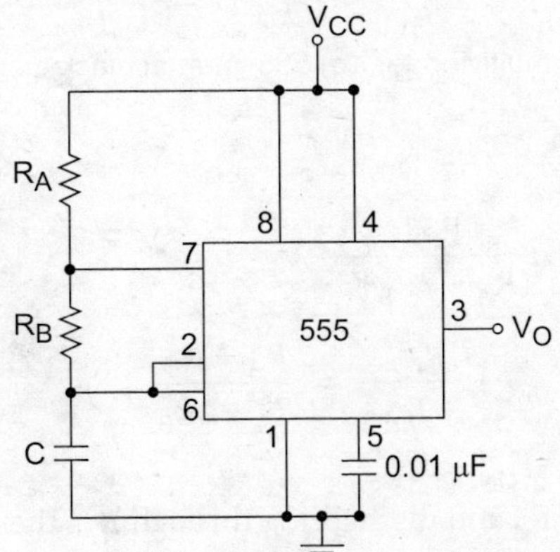

(*a*)

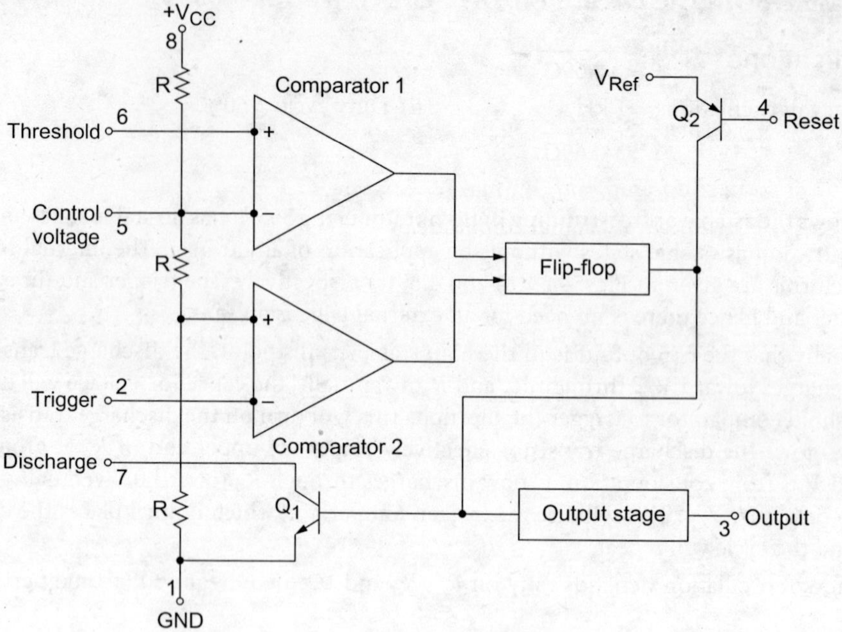

(b)

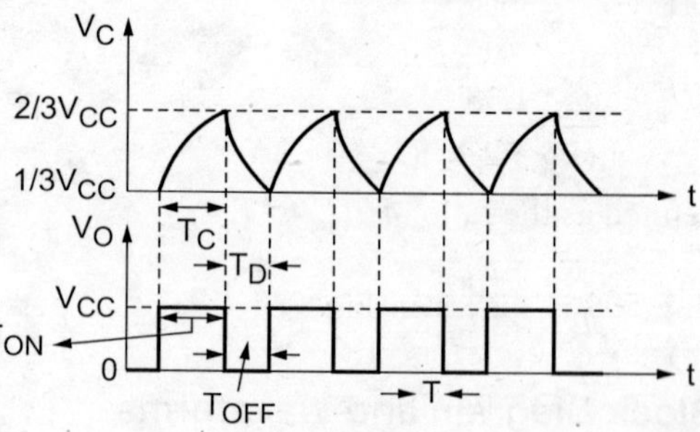

(c)

Fig. 4.39: (a) Asymmetrical astable multivibrator, (b) 555 internal block diagram, and (c) voltage waveforms

Design Equations

The frequency of oscillations, $f_O = \dfrac{1.38}{(R_A + 2R_B)C}$

Duty cycle, $D = \dfrac{T_{ON}}{T_{ON} + T_{OFF}} = \dfrac{R_A + R_B}{R_A + 2R_B}$

Time period, $T = T_{ON} + T_{OFF}$

Charging occurs through R_A and R_B and discharging through R_B. Therefore,

Charging time, $T_C = 0.69(R_A + R_B)C$

Discharging time, $T_D = 0.69R_BC$

$$R_B = \frac{T_D}{0.69C}$$

$$R_A = \frac{T_C}{0.69C} - R_B$$

The slew rate minimum of op-amp = $\pi f_O V_O(\max)$
where $V_O(\max)$ is the maximum output signal amplitude.

Design Steps

Given, $f_O = 1$ kHz and duty cycle, $D = 60\% = 0.6$

Time period, $T = \frac{1}{f_O} = 1$ ms

Charging time, $T_C = 0.6T = 0.6$ ms

Discharging time, $T_D = 0.4T = 0.4$ ms

Assume $C = 0.1$ μF.

Then, $R_B = 5.8$ kΩ

$R_A = 2.9$ kΩ

4.40 MONOSTABLE MULTIVIBRATOR USING 555

Design Specification

Frequency of oscillations $f_o = 1$ kHz.

Theory

The monostable multivibrator using 555 is shown in Fig. 4.40(*a*). The internal block diagram and various waveforms are given in Figs. 4.40(*b*) and 4.40(*c*) respectively. It has one stable state and one quasi-stable state. In this circuit, the 555 will produce an output pulse of predetermined width when a negative going pulse is applied to the trigger input. When the output is low, the circuit is said to be in stable state. The three input components, R_T, C_T, and D serve to limit and differentiate the applied pulse. In this way, a very narrow pulse will result which reduces the possibility of false triggers.

Assume that the output of the 555 is initially low. This implies that the discharge transistor Q_1 is on, shorting the timing capacitor C. Upon application of trigger pulse to pin 2, the output of the trigger comparator changes state, firing the flip-flop, which in turn causes the output to go high and turn off the discharge transistor. At this point, C begins to charge toward V_{CC} through R. When the capacitor voltage reaches $2/3V_{CC}$, the output of the threshold comparator becomes high, the flip-flop will reset, and output goes low as shown in figure. This turns on the discharge transistor, discharging the timing capacitor, and the circuit is ready to accept a new input pulse. Without the input wave shaping network, the trigger pulse must be narrower than the desired output pulse. The output pulse width is given by

$$T = 1.1RC$$

Circuit Diagram, Internal Block Diagram and Waveforms

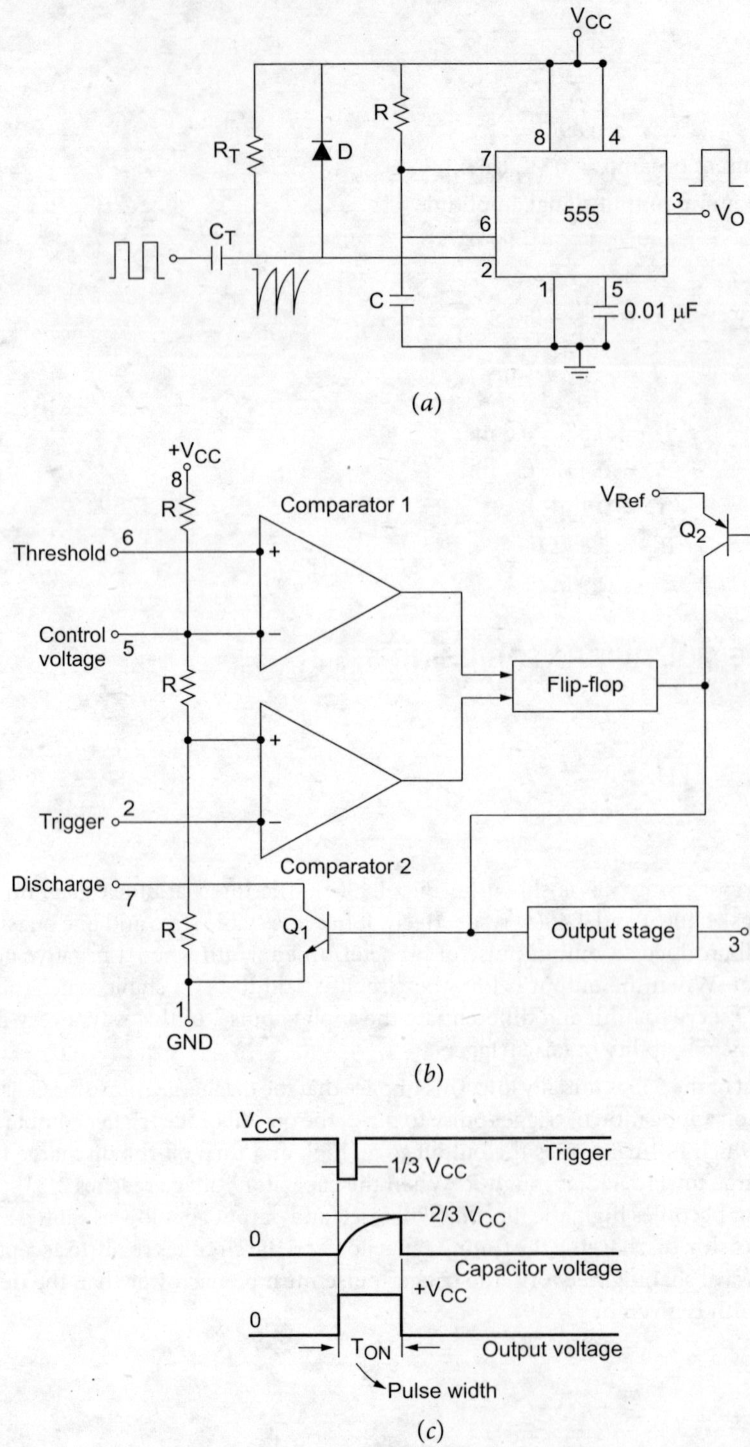

Fig. 4.40: (a) Monostable multivibrator, (b) 555 internal block diagram, and (c) voltage waveforms

Design Equations

The frequency of oscillations, $f_O = \dfrac{1}{T} = \dfrac{1}{1.1RC}$

where T is the time period.

The trigger pulse width, $t_p = R_T C_T << T$

The slew rate minimum of op-amp $= \pi f_O V_O(\text{max})$

where $V_O(\text{max})$ is the maximum output signal amplitude.

Design Steps

Given, $f_O = 1$ kHz

$T = 1 \text{ ms} = 1.1RC$

Let $C = 0.1$ μF. Then $R = 9.1$ kΩ

The trigger pulse width, $t_p << T$

Let $t_p = \dfrac{T}{100} = 0.01$ ms

$R_T C_T = 0.01$ ms

Choose $C_T = 0.01$ μF. Then $R_T = 1$ kΩ

Choose 1N4001 diode for D.

Choose 741 op-amp with saturation voltages of $V_{sat} = \pm 13$ V

Slew rate (min.) = 0.082 V/μs

This value is within the rated value 0.5 V/μs of 741 op-amp. Hence 741 op-amp is suitable for the present design.

4.41 R-2R LADDER 4-BIT DIGITAL-TO-ANALOG CONVERTER (DAC)

Theory

The *R*-2*R* ladder type digital-to-analog converter is shown in Fig. 4.43. It is called an *R*-2*R* ladder *D*/*A* converter, since the input network resembles the rungs on a ladder and the resistors in the input network are either equal (*R*) or have a 2 : 1 ratio (2*R*). One advantage of the *R*-2*R* converter is that the resistors have a 2 : 1 ratio regardless of the number of bits being converted. This makes matching resistors much easier and even makes the use of integrated resistors practical.

The operation of the circuit can be understood by Thevenizing the input circuit for one or more digital input numbers. Once the input circuit has been simplified with Thevenin's theorem, the input voltage to a simple inverting amplifier circuit is the Thevenin equivalent voltage and its gain is determined by the ratio of feedback resistance to Thevenin equivalent input resistance. Here, the least significant input (b_1) produces the least effect on output voltage, and the next input (b_2) has twice as much effect on output voltage. Similarly, bit b_3 has twice the effect of b_2 but only half the effect on output voltage of b_4. These variable effects are identical to the relative weights of the digits in a binary number.

The linearity of the *R*-2*R* ladder *D*/*A* converter circuit is certainly less than optimum. The linearity could be greatly improved by using precision resistors and by driving the digital inputs via analog switches, rather than directly from the output of a digital counter.

Circuit Diagram

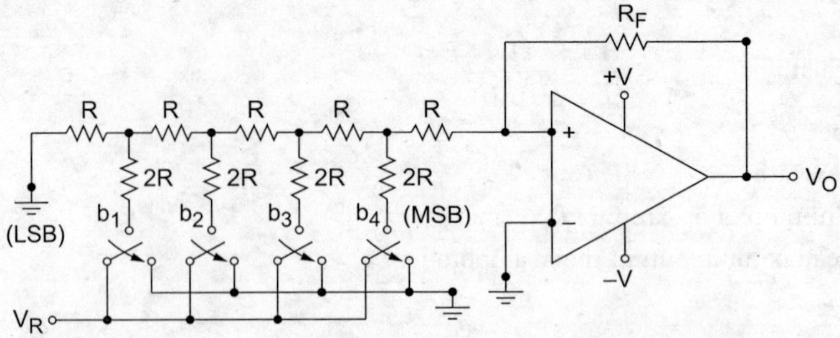

Fig. 4.41: R-2R ladder 4-bit digital-to-analog converter (DAC) using 741 op-amp

Design Equations

The output voltage,
$$V_O = -V_R\left(\frac{R_F}{2R}\right)\cdot\left(\frac{b_4}{2^1}+\frac{b_3}{2^2}+\frac{b_2}{2^3}+\frac{b_1}{2^4}\right)$$

where b_1 is LSB bit and b_4 is MSB bit.

$$\text{Resolution} = \frac{V_{FS}}{2^n - 1},\ V_{FS} = \text{Full-scale output voltage and } n = \text{No. of digital inputs.}$$

Resolution is also given as $\frac{V_R}{2^n}$.

Design Steps

Choose $V_R = 5$ V and $R = 1$ kΩ

$$2R = 2\text{ k}\Omega \quad \text{and} \quad R_F = 2R = 2\text{ k}\Omega$$

$$V_{FS} = 5(1)\left(\frac{1}{2}+\frac{1}{4}+\frac{1}{8}+\frac{1}{16}\right) = 4.6875\text{ V}$$

Resolution = 0.3125 V and is the value of LSB bit.

4.42 FIXED OUTPUT VOLTAGE REGULATOR USING IC723

Design Specifications

(*i*) Output voltage = 15 V (*ii*) Current limit = 100 mA.

Theory

The IC723 can be used to provide high and low positive regulated voltages. It has short circuit protection. The input voltage of IC723 vary from 9.5 to 40 V and provide output voltages from 2 to 37 V. Fig. 4.44 shows the voltage regulator to provide a fixed output voltage of 15 V. The output current can be boosted to provide 5 A or more. For outputs greater than the 7.15 V reference, resistors R_1 and R_2 are used to set the gain of the internal amplifier, *i.e.*, the amount by which the reference will be multiplied. Resistor R_3 simply serves as DC bias compensation for R_1 and R_2. For outputs less than the reference, R_1 and R_2 serve as a voltage divider, which effectively reduces the reference, whereas the amplifier operates with a gain of unity. Again, R_3 serves as bias compensation. In short, the output voltage is once again determined by the reference voltage in conjunction with a pair of resistors. In both cases, the output current limit is set by

$$I_{limit} = \frac{V_{sense}}{R_{sc}}$$

where V_{sense} is approximately 0.65 V at room temperature. Resistor R_{sc} is placed in series with the load, and thus, the current through it is the load current, ignoring the small current required by the adjustment resistors. The current limit transistor is connected so that the voltage across R_{sc} is applied to this transistor's base-emitter junction. The collector of the current limit transistor is connected to the base of the output pass transistor. If the output current rises to the point where the voltage across R_{sc} exceeds approximately 0.65 V, the current limit transistor will turn on, thus shunting output drive current away from the output pass transistor.

Circuit Diagram

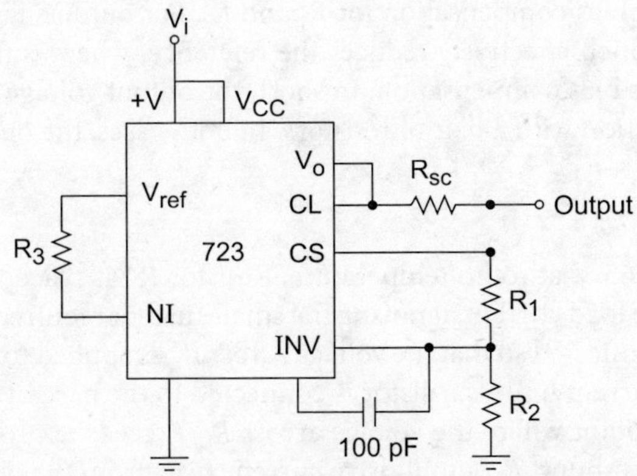

Fig. 4.42: Fixed output voltage regulator using IC723

Design Equations

The reference voltage, $V_{ref} = 7.15$ V

The output voltage, $V_O = V_{ref} \cdot \frac{R_1 + R_2}{R_2}$

The resistor R_1 is given by

$$R_1 = \left(V_O \cdot \frac{R_2}{V_{ref}}\right) - R_2$$

The limiting current, $I_{limit} = \frac{V_{sense}}{R_{sc}}$, where $V_{sense} = 0.65$ V at room temperature and R_{sc} is current sense resistor.

For minimum temperature drift, R_3 is included in the circuit and is chosen as

$$R_3 = R_1 \,||\, R_2$$

Design Steps

Choosing $R_2 = 10$ kΩ, R_1 can be found as

$$R_1 = 10.98 \text{ k}\Omega \quad \text{and} \quad R_{sc} = 6.5\ \Omega$$

$$R_3 = 5.23 \text{ k}\Omega$$

4.43 ADJUSTABLE OUTPUT VOLTAGE REGULATOR USING IC723

Design Specifications

(*i*) Continuously adjustable output voltage = 2 to 6 V

(*ii*) Current limit = 1 A.

Theory

The IC723 can be used to provide high and low positive regulated voltages. It has short circuit protection. The input voltage of IC723 vary from 9.5 to 40 V and provide output voltages from 2 to 37 V. Fig. 4.45 shows a voltage regulator using IC723 to provide a continuously adjustable output voltage of 2 to 6 V. The output current can be boosted to provide 5 A or more. For outputs greater than the 7.15 V reference, resistors R_1 and R_2 are used to set the gain of the internal amplifier, *i.e.*, the amount by which the reference will be multiplied. Resistor R_3 simply serves as DC bias compensation for R_1 and R_2. For outputs less than the reference, R_1 and R_2 serve as a voltage divider, which effectively reduces the reference, whereas the amplifier operates with a gain of unity. Again, R_3 serves as bias compensation. In short, the output voltage is once again determined by the reference voltage in conjunction with a pair of resistors. In both cases, the output current limit is set by

$$I_{\text{limit}} = \frac{V_{\text{sense}}}{R_{sc}}$$

where V_{sense} is approximately 0.65 V at room temperature. Resistor R_{sc} is placed in series with the load, and thus the current through it is the load current, ignoring the small current required by the adjustment resistors. The current limit transistor is connected so that the voltage across R_{sc} is applied to this transistor's base-emitter junction. The collector of the current limit transistor is connected to the base of the output pass transistor. If the output current rises to the point where the voltage across R_{sc} exceeds approximately 0.65 V, the current limit transistor will turn on, thus shunting output drive current away from the output pass transistor.

In order to produce 1 A load current, an external pass transistor will be used. To obtain the desired voltage adjustment, resistor R_1 is replaced with a series potentiometer and resistor combination (R_{1a}, R_{1b}). The minimum value for R_1 will be R_{1b}, and the maximum value will be ($R_{1a} + R_{1b}$).

Circuit Diagram

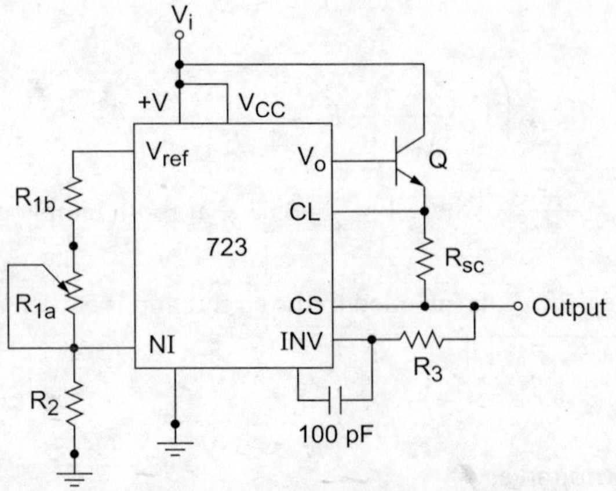

Fig. 4.43: Continuously adjustable output voltage regulator using IC723

Design Equations

The reference voltage, $V_{ref} = 7.15$ V

The output voltage, $V_O = V_{ref} \cdot \dfrac{R_2}{R_1 + R_2}$

For the maximum value of V_O,

$$\frac{V_{ref}}{V_O} = \frac{R_{1b} + R_2}{R_2}$$

For the minimum value of V_O,

$$\frac{V_{ref}}{V_O} = \frac{R_{1a} + R_{1b} + R_2}{R_2}$$

The limiting current, $I_{limit} = \dfrac{V_{sense}}{R_{sc}}$, where $V_{sense} = 0.65$ V at room temperature and R_{sc} is current sense resistor.

For minimum temperature drift, R_3 is included in the circuit and is chosen as $R_3 = R_1 \,||\, R_2$

Design Steps

For $V_O = 6$ V, $\dfrac{R_{1b} + R_2}{R_2} = \dfrac{7.15}{6} = 1.192$

If R_2 is assumed as unity, then the ratio of $(R_{1b} + R_2)$ to $R_2 = 1.192 : 1$ or the ratio of R_{1b} to $R_2 = 0.192 : 1$

For $V_O = 2$ V, $\dfrac{R_{1a} + R_{1b} + R_2}{R_2} = 3.575$

Then the ratio of $(R_{1a} + R_{1b})$ to $R_2 = 2.575 : 1$

Since the ratio of R_{1b} to $R_2 = 0.192 : 1$, the ratio of R_{1a} to R_2 will be 2.383 : 1

Choosing $R_{1a} = 10$ kΩ,

$$R_2 = 4.196 \text{ k}\Omega$$

$$R_{1b} = 0.806 \text{ k}\Omega$$

$$R_{sc} = 0.65\ \Omega$$

$$R_3 = R_1 \,||\, R_2$$

Since R_1 is adjustable, a mid-point value of R_{1a} and R_{1b} can be chosen.

$$R_1 = 5.403 \text{ k}\Omega$$

$$R_3 = 2.362 \text{ k}\Omega$$

The external pass transistor Q must be able to handle the current of 1 A continuously. Also, a minimum β specification is required. As the LM723 will be driving the pass transistor, the LM723 only needs to produce base drive current. With a maximum output of 150 mA, this translates to a minimum β of

$$\beta_{min} = \frac{I_c}{I_b} = \frac{1}{0.15} = 6.67$$

Chose a power transistor such as 2N3019 that satisfies the I_C and β values.

4.44 5V OUTPUT VOLTAGE REGULATOR

Design Specifications

(*i*) Output voltage = 5 V (*ii*) Output current = 1 A.

Theory

Voltage sources in a circuit may have fluctuations resulting in not providing fixed voltage outputs. A voltage regulator IC maintains the output voltage at a constant value. IC 7805, a member of 78xx series of fixed linear voltage regulators used to maintain such fluctuations, is a popular voltage regulator integrated circuit. 7805 IC provides +5 volts regulated output voltage with a provision to add heat sink and has a current rating of up to 1.5 A. A voltage regulator that provides a constant 5 V output voltage using 7805 IC is shown in Fig. 4.46. It has 3 terminals, INPUT, GND, and OUTPUT. It accepts an input voltage of 7 to 35 V and provides an output of 4.8 to 5.2 V. The dropout voltage between input and out terminals is typically 2 V.

The significant difference between the input voltage and the output voltage is released as heat. The greater the difference between the input and output voltage, more the heat generated. If the regulator does not have a heat sink to dissipate this heat, it can get destroyed and malfunction. Hence, it is advisable to limit the input voltage to a maximum of 2-3 volts above the output voltage.

It is recommended that the regulator input be bypassed with capacitor if the regulator is connected to the power supply filter with long lengths, or if the output load capacitance is large. An input bypass capacitor should be selected to provide good high frequency characteristics to ensure stable operation under all load conditions. A 0.33 μF or larger tantalum, mylar or other capacitor having low internal impedance at high frequencies should be chosen. The bypass capacitor should be mounted with the shortest possible leads directly across the regulators input terminals. In order to improve the transient response, a 0.1 μF capacitor is connected across the output terminals.

Circuit Diagram

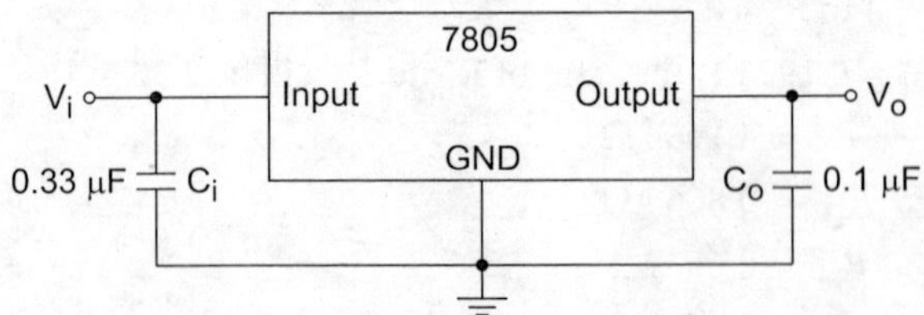

Fig. 4.44: 5 V voltage regulator using IC 7805

Design Equations

$$V_i \geq V_o + V_{DO}$$

where V_{DO} is the dropout voltage for the 7805.

Design Steps

To get a regulated output voltage of + 5 V with 1 A output current, choose voltage regulator IC 7805.

The minimum input voltage $V_i = V_o + V_{DO} = 5 + 2 = 7$ V.

Choose $V_i = 9$ V.

The input and output filter capacitors for the regulator circuit are $C_i = 0.33$ μF and $C_o = 0.1$ μF.

4.45 DUAL-POLARITY VOLTAGE REGULATOR

Design Specifications

(*i*) Output voltage = ± 15 V (*ii*) Output current = 1 A.

Theory

Voltage sources in a circuit may have fluctuations resulting in not providing fixed voltage outputs. A voltage regulator IC maintains the output voltage at a constant value. A voltage regulator that provides output voltages of ± 15 V using 7815 and 7915 ICs is shown in Fig. 4.47. IC 7815, a member of 78xx series of fixed linear voltage regulators used to maintain such fluctuations, is a popular voltage regulator integrated circuit. 7815 IC provides + 15 volts regulated output voltage with a provision to add heat sink and has a current rating of up to 1.5 A. It has 3 terminals, INPUT, GND, and OUTPUT. It accepts an input voltage of 7 to 35 V and provides an output of 14.4 to 15.6 V. The dropout voltage between input and out terminals is typically 2 V.

The significant difference between the input voltage and the output voltage is released as heat. The greater the difference between the input and output voltage, more the heat generated. If the regulator does not have a heat sink to dissipate this heat, it can get destroyed and malfunction. Hence, it is advisable to limit the input voltage to a maximum of 2 to 3 volts above the output voltage.

It is recommended that the regulator input be bypassed with capacitor if the regulator is connected to the power supply filter with long lengths, or if the output load capacitance is large. An input bypass capacitor should be selected to provide good high frequency characteristics to ensure stable operation under all load conditions. A 0.33 μF or larger tantalum, mylar or other capacitor having low internal impedance at high frequencies should be chosen. The bypass capacitor should be mounted with the shortest possible leads directly across the regulators input terminals. In order to improve the transient response, a 0.1 μF capacitor is connected across the output terminals.

The negative regulated voltages are provided by 79xx series of fixed linear voltage regulators. 7915 IC provides -15 volts regulated output voltage with a provision to add heat sink and has a current rating of up to 1.5 A. It has 3 terminals, INPUT, GND, and OUTPUT. It accepts a typical input voltage of – 23 V and provides an output of – 14.4 to – 15.6 V. The dropout voltage between input and out terminals is typically 1.1 V.

Bypass capacitors are necessary for stable operation of the IC 7915 over the input voltage and output current ranges. Output bypass capacitors will improve the transient response by the regulator. The bypass capacitors, 2.2 μF on the input and 1.0 μF on the output should be ceramic or solid tantalum which have good high frequency characteristics. If aluminium electrolytic capacitors are used, their values should be 10 μF or larger. The bypass capacitors should be mounted with the shortest leads, and if possible, directly across the regulator terminals.

Circuit Diagram

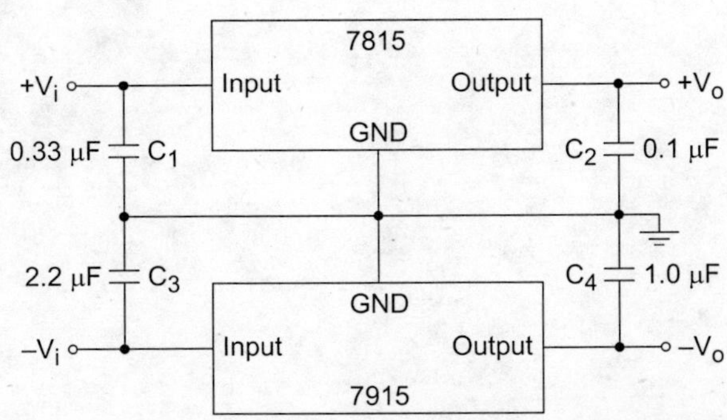

Fig. 4.45 ± 15 V voltage regulator using ICs 7815 and 7915

Design Equations

$$|V_i| \geq |V_o| + V_{DO}$$

where V_{DO} is the dropout voltage for the IC.

Design Steps

To get a regulated output voltage of + 15 V with 1 A output current, choose voltage regulator IC 7815. The minimum input voltage $V_i = V_o + V_{DO} = 15 + 2 = 17$ V. So, choose $V_i = 18$ V. The input and output filter capacitors for the regulator are

$$C_1 = 0.33\ \mu\text{F} \quad \text{and} \quad C_2 = 0.1\ \mu\text{F}$$

Similarly, to get a regulated output voltage of – 15 V with 1 A output current, choose voltage regulator IC 7915. Since $V_{DO} = 1.1$ V, the minimum input voltage $V_i = -16.1$ V. So, choose $V_i = -18$ V. The input and output filter capacitors for the regulator are

$$C_3 = 2.2\ \mu\text{F} \quad \text{and} \quad C_4 = 1.0\ \mu\text{F}$$

❑❑❑

APPENDIX A: DIODES DATASHEETS

A.1 1N914/A/B HIGH-SPEED SWITCHING DIODES

Features

- High switching speed: max 4 ns
- Continuous reverse voltage: max 75 V
- Repetitive peak reverse voltage: max 100 V
- Repetitive peak forward current: max 225 mA

Maximum Ratings and Thermal Characteristics

(Rating at 25°C ambient temperature unless otherwise specified)

Parameter			Symbol	Value	Unit
Non-Repetitive Peak Reverse voltage			V_{RM}	100	V
Reverse Voltage			V_R	75	V
Average Rectified Forward Current		1N9I4	$I_{F(AV)}$	75	mA
		1N914 A/B		200	
Forward Continuous Current		1N914	I_{FM}	150	mA
		IN914 A/B		300	
Non-repetitive Peak Forward Surge Current		at t = 1 s	I_{FSM}	1	A
	IN914	at t = 1 μs		1	
	IN914 A/B	at t = 1 μs		4	
Power Dissipation			P_{tot}	500	mW
Operating and Storage Temperature Range			T_J, T_{stg}	–65 to +175	°C

Electrical Characteristics

(T_J = 25°C unless otherwise noted)

Parameter		Symbol	Test Condition	Min	Typ	Max	Unit
Reverse Current		I_R	V_R = 20 V	—	—	25	nA
			V_R = 75 V	—	—	5	μA
			V_R = 20 V, Tj = 150°	—	—	50	μA
Forward Voltage	1N914	V_F	I_F = 10 mA	—	—	1.0	V
	1N914A		I_F = 20 mA	—	—	1.0	V
	1N914B		I_F = 5 mA	0.62	—	0.72	V
	1n914B		I_F = 100 mA	—	—	1.0	V
Diode Capacitance		C_d	f = 1 MHz; V_R = 0	—	—	4.0	pF
Reverse Recovery Time		Trr	I_F = 10 mA to I_R = 1 mA V_R = 1 mA, R_L = 100 Ω	—	—	4	ns

A.2 1N4001 GENERAL PURPOSE PLASTIC RECTIFIER

Features

- Low forward voltage drop
- Low leakage current
- High forward surge capability

Typical Applications

For use in general purpose rectification of power supplies, inverters, converters, and freewheeling diodes application.

Primary Characteristics

$I_{F(AV)}$	1.0 A
V_{RRM}	50 V
I_{FSM} (8.3 ms sine-wave)	30 A
I_{FSM} (square wave t_p = 1 ms)	45 A
V_F	1.1 V
I_R	5.0 μA
T_J max.	150°C
Package	DO-41
Circuit configuration	Single

Maximum Ratings

(T_A = 25°C unless otherwise noted)

Parameter		Symbol	Value
Maximum repetitive peak reverse voltage		V_{RRM}	50 V
Maximum RMS voltage		V_{RMS}	35 V
Maximum DC blocking voltage		V_{DC}	50 V
Maximum average forward rectified current 0.375″ (9.5 mm) lead length at T_A = 75°C		$I_{F(AV)}$	1.0 A
Peak forward surge current 8.3 ms single half sine-wave superimposed on rated load		I_{FSM}	30 A
Non-repetitive peak forward surge current square waveform T_A = 25°C	t_p = 1 ms t_p = 2 ms t_p = 5 ms	I_{FSM}	45 A 35 A 30 A
Maximum full load reverse current, full cycle average 0.375″ (9.5 mm) lead length T_L = 75°C		$I_{R(AV)}$	30 μA
Rating for fusing (t < 8.3 ms)		I^2t	3.7 A^2s
Operating junction and storage temperature range		T_J, T_{STG}	– 50 to + 150°C

A.3 1N4148 SMALL SIGNAL FAST SWITCHING DIODE

Applications

- Extreme fast switches

Absolute Maximum Ratings

(T_{amb} = 25°C, unless otherwise specified)

Parameter	Test Condition	Symbol	Value	Unit
Repetitive peak reverse voltage		V_{RRM}	100	V
Reverse voltage		V_R	75	V
Peak forward surge current	t_p = 1 μs	I_{FSM}	2	A
Repetitive peak toward current		I_{FRM}	500	mA
Forward continuous current		I_F	300	mA
Average forward current	$V_R = 0$	$I_{F(AV)}$	150	mA
Power dissipation	l = 4 mm, T_L = 45°C	P_{tot}	440	mW
	l = 4 mm, $T_L \leq$ 25°C	P_{tot}	500	mW

Electrical Characteristics

(T_{amb} = 25°C, unless otherwise specified)

Parameter	Test Condition	Symbol	Min.	Typ.	Max.	Unit
Forward voltage	I_F = 10 mA	V_F			1	V
Reverse current	V_R = 20 V	I_R			25	nA
	V_R = 20 V, T_j = 150°C	I_R			50	μA
	V_R = 75 V	I_R			5	μA
Breakdown voltage	I_R = 100 mA, t_p/T = 0.01, t_p = 0.3 ms	$V_{(BR)}$	100			V
Diode capacitance	V_R = 0 V, f = 1 MHz, V_{HF} = 50 mV	C_D			4	pF
Rectification efficiency	V_{HF} = 2 V, f = 100 MHz	η_r	45			%
Reverse recovery time	$I_F = I_R$ = 10 mA, I_R = 1 mA	t_{rr}			8	ns
	I_F = 10 mA, V_R = 6 V, $I_R = 0.1 \times I_R$, R_L = 100 Ω	t_{rr}			4	ns

A.4 IR325 5 mm INFRARED LED

Features

- High reliability
- High radiant intensity
- Peak wavelength $\lambda p = 940$ nm
- 2.54 mm lead spacing
- Low forward voltage

Applications

- Free air transmission system
- Infrared remote control units with high power requirement
- Smoke detector
- Infrared applied system

Device Selection Guide

LED Part No.	Chip Material	Lens Colour
IR	GaAlAs	Blue

Absolute Maximum Ratings

($T_A = 25°C$)

Parameter	Symbol	Rating	Units
Continuous forward current	I_F	100	mA
Peak forward current	I_{FP}	1.0	A
Reverse voltage	V_R	5	V
Operating temperature	T_{opr}	– 40 ~ + 85	°C
Storage temperature	T_{stg}	– 40 ~ + 85	°C
Soldering temperature	T_{sol}	260	°C
Power dissipation at (or below) 25°C free air temperature	P_d	150	mW

A.5 BPV10NF SILICON PIN PHOTODIODE

Features

- Package type: leaded
- Package form: $T\text{-}1\frac{3}{4}$
- Dimensions (in mm): ϕ 5
- Radiant sensitive area (in mm^2): 0.78
- Leads with stand-off
- High radiant sensitivity
- Daylight blocking filter matched with 870 nm to 950 nm emitters
- High bandwidth: > 100 MHz at $V_R = 12$ V
- Fast response times
- Angle of half sensitivity: $\phi = \pm 20°$

Applications

- High speed detector for infrared radiation
- Infrared remote control and free air data transmission systems, e.g., in combination with TSFFxxxx series IR emitters

Product Summary

Component	I_{ra} (μA)	ϕ(deg)	$\lambda_{0.5}$ (nm)
BPV10NF	60	± 20	790 to 1050

Absolute Maximum Ratings

(T_{amb} = 25°C, unless otherwise specified)

Parameter	Test Condition	Symbol	Value	Unit
Reverse voltage		V_R	60	V
Power dissipation	$T_{amb} \leq 25°C$	P_V	215	mW
Junction temperature		T_j	100	°C
Operating temperature range		T_{amb}	–40 to +100	°C
Storage temperature range		T_{stg}	–40 to +100	°C
Soldering temperature	$t \leq 5$ s, 2 mm from body	T_{sd}	260	°C
Thermal resistance junction/ambient	Connected with Cu wire, 0.14 mm^2	R_{thJA}	350	K/W

Basic Characteristics

(T_{amb} = 25°C, unless otherwise specified)

Parameter	Test Condition	Symbol	Min.	Typ.	Max.	Unit
Forward voltage	I_F = 50 mA	V_F		1.0	1.3	V
Breakdown voltage	I_R = 100 μA, E = 0	$V_{(BR)}$	60			V
Reverse dark current	V_R = 20 V, E = 0	I_{ro}		1	5	nA
Diode capacitance	V_R = 0 V, f = 1 MHz, E = 0	C_D		11		pF
Open circuit voltage	E_o = 1 mW/cm², λ = 870 nm	V_O		450		mV
Short circuit current	E_o = 1 mW/cm², λ = 870 nm	I_K		50		μA
Reverse light current	E_o = 1 mW/cm², λ = 870 nm, V_R = 5 V	I_{ra}	30	55		μA
	E_o = 1 mW/cm², λ = 870 nm, V_R = 5 V	I_{ra}		60		μA
Temperature coefficient of I_{ra}	E_o = 1 mW/cm², λ = 870 nm, V_R = 5 V	TK_{Ira}		–0.1		%/K
Absolute spectral sensitivity	V_R = 5 V, λ = 870 nm	s(λ)		0.55		A/W
Angle of half sensitivity		ϕ		±20		deg
Wavelength of peak sensitivity		λ_p		940		nm
Range of spectral bandwidth		$\lambda_{0.5}$		790 to 1050		nm
Quantum efficiency	λ = 950 nm	η		70		%
Noise equivalent power	V_R = 20 V, λ = 950 nm	NEP		3×10^{-14}		W/√Hz
Detectivity	V_R = 20 V, λ = 950 nm	D*		3×10^{12}		cm√Hz/W
Rise time	V_R = 50 V, R_L = 50 Ω, λ = 820 nm	t_r		2.5		ns
Fall time	V_R = 50 V, R_L = 50 Ω, λ = 820 nm	t_f		2.5		ns

A.6 1N753 6.2V 0.5W SILICON PLANAR ZENER DIODE

Features

- Silicon Planar Power Zener Diodes
- Standard Zener voltage tolerance is ±5% for "A" suffix.

Maximum Ratings

At 25°C ambient temperature unless otherwise specified.

	Symbol	Value	Unit
Power dissipation at T_L 75°C	P_{tot}	500[(1)]	mW
Maximum junction temperature	T_j	175	°C
Storage temperature range	T_S	–65 to +175	°C

	Symbol	Min.	Typ.	Max.	Unit
Thermal resistance junction to ambient air	R_{thJA}	—	—	300[(1)]	°C/W
Forward voltage at I_F = 200 mA	V_F	—	—	1.5	Volts

Electrical Characteristics

At 25°C ambient temperature unless specified

Nominal Zener Voltage V_Z @ I_{ZT}[(3)] (Volts)	Test Current I_{ZT} (mA)	Maximum Zener Impedance Z_{ZT} @ I_{ZT}[(1)] (Ω)	Maximum Regulator Current I_{ZM}[(2)] (mA)	Maximum Reverse Leakage Current T_A = 25°C I_R @ V_R = 1V (μA)	Maximum Reverse Leakage Current T_A = 150°C I_R @ V_R = 1V (μA)
6.2	20	7	60	0.1	20

Notes

(1) The Zener impedance is derived from the 1 KHz AC voltage which results when an AC current having an RMS value equal to 10% of the Zener current (I_{ZT}) is superimposed on I_{ZT}. Zener impedance is measured at two points to insure a sharp knee on the breakdown curve and to eliminate unstable units.

(2) Valid provided that leads at a distance of 3/8″ from case are kept at ambient temperature

(3) Measured with device junction in thermal equilibrium

A.7 1N4733A 5.1V 1W ZENER DIODE

Features

- Silicon planar power Zener diodes
- For use in stabilizing and clipping circuits with high power rating
- Standard Zener voltage tolerance is ± 5 %

Applications

- Voltage stabilization

Absolute Maximum Ratings

(T_{amb} = 25°C, unless otherwise specified)

Parameter	Test Condition	Symbol	Value	Unit
Power dissipation	Valid provided that leads at a distance of 4 mm from case are kept at ambient temperature	P_{tot}	1000	mW
Zener current		I_Z	P_V/V_Z	mA
Thermal resistance junction to ambient air	Valid provided that leads at a distance of 4 mm from case are kept at ambient temperature	R_{thJA}	110	K/W
Junction temperature		T_j	175	°C
Storage temperature range		T_{stg}	–65 to +175	°C
Forward voltage (max.)	I_F = 200 mA	V_F	1.2	V

Electrical Characteristics

(T_{amb} = 25°C, unless otherwise specified)

Part Number	Zener Voltage Range	Test Current		Reverse Leakage Current		Dynamic Resistance f = 1 kHz		Surge Current	Regulator Current
	V_Z at I_{ZT1}	I_{ZT1}	I_{ZT2}	I_R at V_R		Z_{ZT} at I_{ZT1}	Z_{ZK} at I_{ZT2}	I_R	I_{ZM}
	V	mA	mA	μA	V	Ω		mA	mA
	Nom.			Max.		Typ.	Max.		Max.
1N4733A	5.1	49	1	10	1	7	550	890	178

A.8 1N4737A 7.5V 1W ZENER DIODE

Features

- 1.0 Watt Power Dissipation
- Standard V_Z Tolerance is 5%
- Polarity: Cathode Band

Maximum Ratings

@T_A = 25°C unless otherwise specified

Characteristic	Symbol	Value	Unit
Zener current	I_Z	Pd/V_Z	mA
Power dissipation derate above 50°C	P_d	1.0 6.67	W mW/°C
Thermal Resistance - Junction to ambient air	$R_{\theta JA}$	175	°C/W
Forward voltage @ I_F = 200 mA	V_F	1.2	V
Operating and storage temperature range	T_j, T_{STG}	–65 to +175	°C

Notes

1. Valid provided that leads are kept at T_L @ 50°C with lead length = 9.5 mm (3/8″) from case.

Electrical Characteristics

@T_A = 25°C unless otherwise specified

Part Number	Nominal Zener Voltage	Test Current	Maximum Zener Impedance			Maximum Reverse Leakage Current		Max Surge Current 8.3 ms	Temperature Coefficient @ I_{ZT}
	V_Z @ I_{ZT}	I_{ZT}	Z_{ZT} @ I_{ZT}	Z_{ZK} @ I_{ZK}	I_{ZK}	I_R	@ V_R	I_{ZS}	
	(V)	(mA)	(Ω)	(Ω)	(mA)	(μA)	(V)	(mA)	%/°C
1N4737A	7.5	34	4.0	700	0.5	10	5.0	605	+0.02 to +0.065

A.9 IN4738A 8.2V 1.3W ZENER DIODE

Features

- Silicon planar power Zener diodes
- For use in stabilizing and clipping circuits with high power rating
- Standard Zener voltage tolerance is ± 5 %

Applications

- Voltage stabilization

Absolute Maximum Ratings

(T_{amb} = 25°C, unless otherwise specified)

Parameter	Test Condition	Symbol	Value	Unit
Power dissipation	Valid provided that leads at a distance of 4 mm from case are kept at ambient temperature	P_{tot}	1300	mW
Zener current		I_Z	P_V/V_Z	mA
Thermal resistance junction to ambient air	Valid provided that leads at a distance of 4 mm from case are kept at ambient temperature	R_{thJA}	110	K/W
Junction temperature		T_j	175	°C
Storage temperature range		T_{stg}	–65 to +175	°C
Forward voltage (max.)	I_F = 200 mA	V_F	1.2	V

Electrical Characteristics

(T_{amb} = 25°C, unless otherwise specified)

Zener Voltage Range(1)	Test Current		Reverse Leakage Current		Dynamic Resistance f = 1 kHz		Surge Current(3)	Regulator Current(2)
V_Z at I_{ZT1}	I_{ZT1}	I_{ZT2}	I_R at V_R		Z_{ZT} at I_{ZT1}	Z_{ZK} at I_{ZT2}	I_R	I_{ZM}
V	mA	mA	μA	V	Ω		mA	mA
Nom.			Max.		Typ.	Max.		Max.
8.2	31	0.5	10	6	4.5	700	550	110

Notes

(1) Based on DC measurement at thermal equilibrium while maintaining the lead temperature (T_L) at 30°C + 1°C, 9.5 mm (3/8″) from the diode body

(2) Valid provided that electrodes at a distance of 4 mm from case are kept at ambient temperature

(3) t_p = 10 ms.

A.10 1N5221 2.4V 0.5W SMALL SIGNAL ZENER DIODE

Features

- Silicon planar power Zener diodes
- Standard Zener voltage tolerance is ± 5 % with a "*B*" suffix in the ordering code (*e.g.*, 1N5221B), suffix "*C*" is ± 2 % tolerance

Applications

- Voltage stabilization

Absolute Maximum Ratings

(T_{amb} = 25°C, unless otherwise specified)

Parameter	Test Condition	Symbol	Value	Unit
Power dissipation	$T_L \leq 25°C$	P_{tot}	500	mW
Zener current		I_Z	P_{tot}/V_Z	mA
Thermal resistance junction to ambient air	l = 4 mm, T_L = constant	R_{thJA}	300	K/W
Junction temperature		T_j	175	°C
Storage temperature range		T_{stg}	–65 to +175	°C
Forward voltage (max.)	I_F = 200 mA	V_F	1.1	V

Electrical Characteristics

(T_{amb} = 25°C, unless otherwise specified)

Part Number	Zener Voltage Range	Test Current		Reverse Leakage Current		Dynamic Resistance f = 1 kHz		Temperature Coefficient
	V_Z at I_{ZT1}	I_{ZT1}	I_{ZT2}	I_R at V_R		Z_{ZT} at I_{ZT1}	Z_{ZK} at I_{ZT2}	α_{VZ}
	V	mA		μA	V	Ω		%/K
	Nom.			Max.		Max.	Max.	Typ.
1N5221	2.4	20	0.25	100	1	30	1200	–0.085

A.11 1N5229 4.3V 0.5W SMALL SIGNAL ZENER DIODE

Features

- Silicon planar power Zener diodes
- Standard Zener voltage tolerance is ± 5 % with a "*B*" suffix in the ordering code (*e.g.*, 1N5229B). suffix "*C*" is ± 2 % tolerance

Applications

- Voltage stabilization

Absolute Maximum Ratings

(T_{amb} = 25°C, unless otherwise specified)

Parameter	Test Condition	Symbol	Value	Unit
Power dissipation	$T_L \leq 25°C$	P_{tot}	500	mW
Zener current		I_Z	P_{tot}/V_Z	mA
Thermal resistance junction to ambient air	l = 4 mm, T_L = constant	R_{thJA}	300	K/W
Junction temperature		T_j	175	°C
Storage temperature range		T_{stg}	–65 to +175	°C
Forward voltage (max.)	I_F = 200 mA	V_F	1.1	V

Electrical Characteristics

(T_{amb} = 25°C, unless otherwise specified)

Part Number	Zener Voltage Range	Test Current		Reverse Leakage Current		Dynamic Resistance f = 1 kHz		Temperature Coefficient
	V_Z at I_{ZT1}	I_{ZT1}	I_{ZT2}	I_R at V_R		Z_{ZT} at I_{ZT1}	Z_{ZK} at I_{ZT2}	α_{VZ}
	V	mA		μA	V	Ω		%/K
	Nom.			Max.		Max.	Max.	Typ.
1N5229	4.3	20	0.25	5	1	22	2000	0.055

A.12 1N5231 5.1V 0.5W SMALL SIGNAL ZENER DIODE

Features

- Silicon planar power Zener diodes
- Standard Zener voltage tolerance is ± 5 % with a "*B*" suffix in the ordering code (*e.g.*, 1N5231B), suffix "*C*" is ± 2 % tolerance

Applications

- Voltage stabilization

Absolute Maximum Ratings

(T_{amb} = 25°C, unless otherwise specified)

Parameter	Test Condition	Symbol	Value	Unit
Power dissipation	$T_L \leq 25°C$	P_{tot}	500	mW
Zener current		I_Z	P_{tot}/V_Z	mA
Thermal resistance junction to ambient air	l = 4 mm, T_L = constant	R_{thJA}	300	K/W
Junction temperature		T_j	175	°C
Storage temperature range		T_{stg}	–65 to +175	°C
Forward voltage (max.)	I_F = 200 mA	V_F	1.1	V

Electrical Characteristics

(T_{amb} = 25°C, unless otherwise specified)

Part Number	Zener Voltage Range	Test Current		Reverse Leakage Current		Dynamic Resistance f = 1 kHz		Temperature Coefficient
	V_Z at I_{ZT1}	I_{ZT1}	I_{ZT2}	I_R at V_R		Z_{ZT} at I_{ZT1}	Z_{ZK} at I_{ZT2}	α_{VZ}
	V	mA		µA	V	Ω		%/K
	Nom.			Max.		Max.	Max.	Typ.
1N5231	5.1	20	0.25	5	2	17	1600	0.03

A.13 1N5236 7.5V 0.5W SMALL SIGNAL ZENER DIODE

Features

- Silicon planar power Zener diodes
- Standard Zener voltage tolerance is ± 5 % with a "*B*" suffix in the ordering code (*e.g.*, 1N5236B). suffix "*C*" is ± 2 % tolerance

Applications

- Voltage stabilization

Absolute Maximum Ratings

(T_{amb} = 25°C, unless otherwise specified)

Parameter	Test Condition	Symbol	Value	Unit
Power dissipation	$T_L \leq 25°C$	P_{tot}	500	mW
Zener current		I_Z	P_{tot}/V_Z	mA
Thermal resistance junction to ambient air	l = 4 mm, T_L = constant	R_{thJA}	300	K/W
Junction temperature		T_j	175	°C
Storage temperature range		T_{stg}	–65 to +175	°C
Forward voltage (max.)	I_F = 200 mA	V_F	1.1	V

Electrical Characteristics

(T_{amb} = 25°C, unless otherwise specified)

Part Number	Zener Voltage Range	Test Current		Reverse Leakage Current		Dynamic Resistance f = 1 kHz		Temperature Coefficient
	V_Z at I_{ZT1}	I_{ZT1}	I_{ZT2}	I_R at V_R		Z_{ZT} at I_{ZT1}	Z_{ZK} at I_{ZT2}	α_{VZ}
	V	mA		μA	V	Ω		%/K
	Nom.			Max.		Max.	Max.	Typ.
1N5236	7.5	20	0.25	3	6	6	500	0.058

APPENDIX B: MOSFET DATASHEET

B.1 2N7000 N-CHANNEL ENHANCEMENT MOSFET

Product Summary

Part Number	$V_{(BR)DSS}$ Min (V)	$r_{DS(on)}$ Max (Ω)	$V_{GS(th)}$ (V)	I_D (A)
2N7000	60	5 @ V_{GS} = 10 V	0.8 to 3	0.2

TO-226AA
(TO-92)

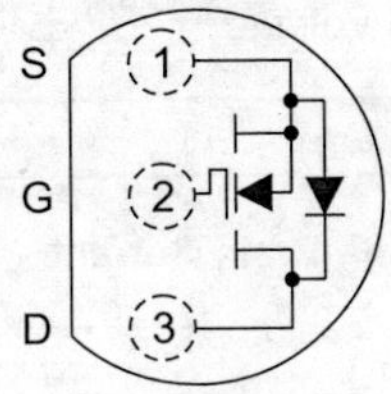

Top View

Features

- Low On-Resistance: 2.5 Ω
- Low Threshold: 2.1 V
- Low Input Capacitance: 22 pF
- Fast Switching Speed: 7 ns
- Low Input and Output Leakage

Benefits

- Low Offset Voltage
- Low-Voltage Operation
- Easily Driven Without Buffer
- High-Speed Circuits
- Low Error Voltage

Applications

- Direct Logic-Level Interface: TTL/CMOS
- Drivers: Relays, Solenoids, Lamps, Hammers, Displays, Memories, Transistors, etc.
- Battery Operated Systems
- Solid-State Relays

Absolute Maximum Ratings

(T_A = 25°C unless otherwise noted)

Parameter		Symbol	2N7000	Unit
Drain-Source Voltage		V_{DS}	60	V
Gate-Source Voltage — Non-Repetitive		V_{GSM}	±40	
Gate-Source Voltage — Continuous		V_{GS}	±20	
Continuous Drain Current	T_A = 25°C	I_D	0.2	A
(T_J = 150°C)	T_A = 100°C		0.13	
Pulsed Drain Current		I_{DM}	0.5	
Power Dissipation	T_A = 25°C	P_D	0.4	W
	T_A = 100°C		0.16	
Thermal Resistance, Junction-to-Ambient		R_{thJA}	312.5	°C/W
Operating Junction and Storage Temperature Range		T_J, T_{stg}	–55 to 150	°C

Specifications

(T_A = 25°C unless otherwise noted)

Parameter	Symbol	Test Condition	Typ	Limits		Unit
				Min	Max	
Static						
Drain-Source Breakdown Voltage	$V_{(BR)DSS}$	V_{GS} = 0 V, I_D = 10 μA	70	60		V
Gate-Threshold Voltage	$V_{GS(th)}$	V_{DS} = V_{GS}, I_D = 1 mA	2.1	0.8	3	
		V_{DS} = V_{GS}, I_D = 0.25 mA	2.0			
Gate-Body Leakage	I_{GSS}	V_{DS} = 0 V, V_{GS} = ± 15 V			±10	nA
Zero Gate Voltage Drain Current	I_{DSS}	V_{DS} = 48 V, V_{GS} = 0 V			1	μA
		T_C = 125°C			1000	
On-State Drain Current	$I_{D(on)}$	V_{DS} = 10 V, V_{GS} = 4.5 V	0.35	0.075		A
		V_{DS} = 7.5 V, V_{GS} = 10 V	1			
Drain-Source On-Resistance	$r_{DS(on)}$	V_{GS} = 4.5 V, I_D = 0.075 A	4.5		5.3	Ω
		V_{GS} = 5 V, I_D = 0.05 A	3.2			
		T_C = 125°C	5.8			
		V_{GS} = 10 V, I_D = 0.5 A	2.4		5	
		T_J = 125°C	4.4		9	
Forward Transconductance	g_{fs}	V_{DS} = 10 V, I_D = 0.2 A		100		mS
Common Source Output Conductance	g_{os}	V_{DS} = 5 V, I_D = 0.05 A	0.5			
Dynamic						
Input Capacitance	C_{iss}	V_{DS} = 25 V, V_{GS} = 0 V f = 1 MHz	22		60	pF
Output Capacitance	C_{oss}		11		25	
Reverse Transfer Capacitance	C_{rss}		2		5	

Specifications

2N7000 (T_A = 25°C unless otherwise noted)

Parameter	Symbol	Test Conditions	Typ	Limits		Unit
				Min	Max	
Switching						
Turn-On Time	t_{ON}	V_{DD} = 15 V, R_L = 25 Ω I_D ≅ 0.5 A, V_{GEN} = 10 V, R_G = 25 Ω	7		10	ns
Turn-Off Time	t_{OFF}		7		10	
Turn-On Time	t_{ON}	V_{DD} = 30 V, R_L = 150 Ω I_D ≅ 0.2 A, V_{GEN} = 10 V, R_G = 25 Ω	7			
Turn-Off Time	t_{OFF}		11			

APPENDIX C: BJTs DATASHEETS

C.1 2N2222 NPN SILICON 0.5W SWITCHING TRANSISTOR

Features

- High current (max. 800 mA)
- Low voltage (max. 40 V).

Applications

- Linear amplification and switching.

Description

NPN switching transistor in a TO-18 metal package. PNP complement: 2N2907A

Quick Reference Data

Symbol	Parameter	Conditions	Min.	Max.	Unit
V_{CBO}	Collector-base voltage	Open emitter	—	60	V
V_{CEO}	Collector-emitter voltage	Open base	—	30	V
I_C	Collector current (DC)		—	800	mA
P_{tot}	Total power dissipation	$T_{amb} \leq 25°C$	—	500	mW
h_{FE}	DC current gain	$I_C = 10$ mA; $V_{CE} = 10$ V	75	—	
f_T	Transition frequency	$I_C = 20$ mA; $V_{CE} = 20$ V, f = 100 MHz	250	—	MHz
t_{off}	Turn-off time	$I_{Con} = 150$ mA; $I_{Bon} = 15$ mA; $I_{Boff} = -15$ mA	—	250	ns

Limiting Values

In accordance with the Absolute Maximum Rating System (IEC 134).

Symbol	Parameter	Conditions	Min.	Max.	Unit
V_{CBO}	Collector-base voltage	Open emitter		60	V
V_{CEO}	Collector-emitter voltage	Open base		30	V
V_{EBO}	Emitter-base voltage	Open collector		5	V
I_C	Collector current (DC)		—	800	mA
I_{CM}	Peak collector current		—	800	mA
I_{BM}	Peak base current		—	200	mA
P_{tot}	Total power dissipation	$T_{amb} \leq 25°C$	—	500	mW
		$T_{case} \leq 25°C$	—	1.2	W
T_{stg}	Storage temperature		–65	+150	°C
T_j	Junction temperature		—	200	°C
T_{amb}	Operating ambient temperature		–65	+150	°C

Thermal Characteristics

Symbol	Parameter	Conditions	Value	Unit
$R_{th\ j\text{-}a}$	Thermal resistance from junction to ambient	in free air	350	K/W
$R_{th\ j\text{-}c}$	Thermal resistance from junction to case		146	K/W

Characteristics

$T_J = 25°C$ unless otherwise specified.

Symbol	Parameter	Conditions	Min.	Max.	Unit
I_{CBO}	Collector cut-off current	$I_E = 0$; $V_{CB} = 50$ V	—	10	nA
		$I_E = 0$; $V_{CB} = 50$ V; $T_{amb} = 150°C$	—	10	μA
I_{EBO}	Emitter cut-off current	$I_C = 0$; $V_{EB} = 3$ V	—	10	nA
h_{FE}	DC current gain	$I_C = 0.1$ mA; $V_{CE} = 10$ V	35	—	
		$I_C = 1$ mA; $V_{CE} = 10$ V	50	—	
		$I_C = 10$ mA; $V_{CE} = 10$ V	75	—	
		$I_C = 150$ mA; $V_{CE} = 1$ V; note 1	50	—	
		$I_C = 150$ mA; $V_{CE} = 10$ V; note 1	100	300	
h_{FE}	DC current gain	$I_C = 500$ mA; $V_{CE} = 10$ V; note 1	30	—	
V_{CEsat}	Collector-emitter saturation voltage	$I_C = 150$ mA; $I_B = 15$ mA; note 1	—	400	mV
		$I_C = 500$ mA; $I_B = 50$ mA; note 1	—	1.6	V
V_{BEsat}	Base-emitter saturation voltage	$I_C = 150$ mA; $I_B = 15$ mA; note 1	—	1.3	V
		$I_C = 500$ mA; $I_B = 50$ mA; note 1	—	2.6	V
C_c	Collector capacitance	$I_E = i_e = 0$; $V_{CB} = 10$ V; f = 1 MHz	—	8	pF
f_T	Transition frequency	$I_C = 20$ mA; $V_{CE} = 20$ V; f = 100 MHz	250	—	MHz

Symbol	Parameter	Conditions	Min.	Max.	Unit
Switching times (between 10% and 90% levels)					
t_{on}	Turn-on time	$I_{Con} = 150$ mA; $I_{Bon} = 15$ mA; $I_{Boff} = -15$ mA	—	35	ns
t_d	Delay time		—	10	ns
t_r	Rise time		—	25	ns
t_{off}	Turn-off time		—	250	ns
t_s	Storage time		—	200	ns
t_f	Fall time		—	60	ns

Note

1. Pulse test: $t_p \leq 300$ μs; $\delta \leq 0.02$.

C.2 2N2222A NPN SILICON 0.5W SMALL SIGNAL SWITCHING TRANSISTOR

Maximum Ratings

(T_A = 25°C unless otherwise noted)

Characteristic	Symbol	Value	Unit
Collector- Emitter Voltage	V_{CEO}	50	Vdc
Collector -Base Voltage	V_{CBO}	75	Vdc
Emitter- Base Voltage	V_{EBO}	6.0	Vdc
Collector Current - Continuous	I_C	800	mAdc
Total Device Dissipation @ T_A = 25°C	P_T	500	mW
Total Device Dissipation @ T_C = 25°C	P_T	1.0	W
Operating and Storage Junction Temperature Range	T_J, T_{stg}	–65 to +200	°C

Thermal Characteristics

Characteristic	Symbol	Max	Unit
Thermal Resistance, Junction to Ambient	$R_{\theta JA}$	325	°C/W
Thermal Resistance, Junction to Case	$R_{\theta JC}$	150	°C/W

Electrical Characteristics

(T_A = 25°C unless otherwise noted)

Characteristic	Symbol	Min.	Max.	Unit
Off Characteristics				
Collector-Emitter Breakdown Voltage	$V_{(BR)CEO}$			
(I_C = 10 mAdc)		50	—	Vdc
Collector-Base Cutoff Current	I_{CBO}			
(V_{CB} = 75 Vdc)		—	10	μAdc
(V_{CB} = 60 Vdc)		—	10	nAdc
Emitter-Base Cutoff Current	I_{EBO}			
(V_{EB} = 6.0 Vdc)		—	10	μAdc
(V_{EB} = 4.0 Vdc)		—	10	nAdc
Collector-Emitter Cutoff Current	I_{CES}			
(V_{CE} = 50 Vdc)		—	50	nAdc
On Characteristics (Note 1)				
DC Current Gain	h_{FE}			—
(I_C = 0.1 mAdc, V_{CE} = 10 Vdc)		50	—	
(I_C = 1.0 mAdc, V_{CE} = 10 Vdc)		75	325	
(I_C = 10 mAdc, V_{CE} = 10 Vdc)		100	—	
(I_C = 150 mAdc, V_{CE} = 10 Vdc)		100	300	
(I_C = 500 mAdc, V_{CE} = 10 Vdc)		30	—	

Contd...

Collector-Emitter Saturation Voltage (I_C = 150 mAdc, I_B = 15 mAdc) (I_C = 500 mAdc, I_B = 50 mAdc)	$V_{CE(sat)}$	 — —	 0.3 1.0	Vdc
Base-Emitter Saturation Voltage (I_C = 150 mAdc, I_B = 15 mAdc) (I_C = 500 mAdc, I_B = 50 mAdc)	$V_{BE(sat)}$	 0.6 —	 1.2 2.0	Vdc
Small-Signal Characteristics				
Magnitude of Small-Signal Current Gain (I_C = 20 mAdc, V_{CE} = 20 Vdc, f = 100 MHz)	$\|h_{fe}\|$	2.5	—	—
Small-Signal Current Gain (I_C = 1.0 mAdc, V_{CE} = 10 Vdc, f = 1 kHz)	h_{fe}	50	—	—
Input Capacitance (V_{EB} = 0.5 Vdc, I_C = 0, 100 kHz ≤ f ≤ 1.0 MHz)	C_{ibo}	—	25	pF
Output Capacitance (V_{CB} = 10 Vdc, I_E = 0, 100 kHz ≤ f ≤ 1.0 MHz)	C_{obo}	—	8.0	pF
Switching (Saturated) Characteristics				
Turn-On Time	t_{on}	—	35	ns
Turn-Off Time	t_{off}	—	300	ns

Product parametric performance is indicated in the Electrical Characteristics for the listed test conditions, unless otherwise noted. Product performance may not be indicated by the Electrical Characteristics if operated under different conditions.

1. Pulse Test: Pulse Width = 300 μs. Duty Cycle ≤ 2.0%.

C.3 2N3019 NPN SILICON LOW POWER TRANSISTOR

Maximum Ratings

(T_A = 25°C unless otherwise noted)

Characteristic	Symbol	Value	Unit
Collector- Emitter Voltage	V_{CEO}	80	Vdc
Collector -Base Voltage	V_{CBO}	140	Vdc
Emitter- Base Voltage	V_{EBO}	7.0	Vdc
Collector Current - Continuous	I_C	1.0	Adc
Total Device Dissipation @ T_A = 25°C	P_T	800	mW
Total Device Dissipation @ T_C = 25°C	P_T	5.0	W
Operating and Storage Junction Temperature Range	T_J, T_{stg}	–65 to +200	°C

Thermal Characteristics

Characteristic	Symbol	Max	Unit
Thermal Resistance, Junction to Ambient	$R_{\theta JA}$	195	°C/W
Thermal Resistance, Junction to Case	$R_{\theta JC}$	30	°C/W

Electrical Characteristics

(T_A = 25°C unless otherwise noted)

Characteristic	Symbol	Min.	Max.	Unit
Off Characteristics				
Collector-Emitter Breakdown Voltage	$V_{(BR)CEO}$			
(I_C = 30 mAdc)		80	—	Vdc
Emitter-Base Cutoff Current	I_{EBO}			
(V_{EB} = 5.0 Vdc)		—	10	nAdc
(V_{EB} = 7.0 Vdc)		—	10	μAdc
Collector-Emitter Cutoff Current	I_{CEO}			
(V_{CE} = 90 Vdc)		—	10	nAdc
Collector-Base Cutoff Current	I_{CBO}			
(V_{CE} = 140 Vdc)		—	10	μAdc
On Characteristics (Note 1)				
DC Current Gain	h_{FE}			—
(I_C = 0.1 mAdc, V_{CE} = 10 Vdc)		50	300	
(I_C = 10 mAdc, V_{CE} = 10 Vdc)		90	—	
(I_C = 150 mAdc, V_{CE} = 10 Vdc)		100	300	
(I_C = 500 mAdc, V_{CE} = 10 Vdc)		50	300	
(I_C = 1.0 Adc, V_{CE} = 10 Vdc)		15	—	

Contd...

Collector-Emitter Saturation Voltage (I_C = 150 mAdc, I_B = 15 mAdc) (I_C = 500 mAdc, I_B = 50 mAdc)	$V_{CE(sat)}$	— —	0.2 0.5	Vdc
Base-Emitter Saturation Voltage (I_C = 150 mAdc, I_B = 15 mAdc)	$V_{BE(sat)}$	—	1.1	Vdc
Small-Signal Characteristics				
Magnitude of Small-Signal Current Gain (I_C = 50 mAdc, V_{CE} = 10 Vdc, f = 20 MHz)	$\|h_{fe}\|$	5.0	20	—
Small-Signal Current Gain (I_C = 1.0 mAdc, V_{CE} = 5 Vdc, f = 1 kHz)	h_{fe}	80	400	—
Output Capacitance (V_{CB} = 10 Vdc, I_E = 0, 100 kHz $\le f \le$ 1.0 MHz)	C_{obo}	—	12	pF
Input Capacitance (V_{EB} = 0.5 Vdc, I_C = 0, 100 kHz $\le f \le$ 1.0 MHz)	C_{ibo}	—	60	pF
Noise Figure (V_{CE} = 10 Vdc, I_C = 100 μAdc, R_g = 1 kΩ, PBW = 200 Hz)	NF	—	4.0	dB
Collector-Base Time Constant (V_{CB} = 10 Vdc, I_C = 10 mAdc, f = 79.8 MHz)	r'_b, C_C	—	400	ps
Switching Characteristics				
Pulse Response	$t_{on} + t_{off}$	—	30	ns

1. Pulse Test: Pulse Width = 300 μs. Duty Cycle ≤ 2.0%.

C.4 2N6040 PNP SILICON MEDIUM POWER TRANSISTOR

Features

- High DC Current Gain – h_{FE} = 2500 (Typ) @ I_C = 4.0 Adc
- Collector-Emitter Sustaining Voltage – @ 100 mAdc

$$V_{CEO(sus)} = 60 \text{ Vdc (Min)}$$

Maximum Ratings

Rating	Symbol	Value	Unit
Collector-Emitter Voltage	V_{CEO}	60	Vdc
Collector-Base Voltage	V_{CB}	60	Vdc
Emitter-Base Voltage	V_{EB}	5.0	Vdc
Collector Current Continuous Peak	I_C	8.0 16	Adc
Base Current	I_B	120	mAdc
Total Power Dissipation @ T_C = 25°C Derate above 25°C	P_D	75 0.60	W W/°C
Operating and Storage Junction Temperature Range	T_J, T_{stg}	–65 to +150	°C

Thermal Characteristics

Characteristic	Symbol	Max	Unit
Thermal Resistance, Junction-to-Case	θ_{JC}	1.67	°C/W
Thermal Resistance, Junction-to-Ambient	θ_{JA}	57	°C/W

Electrical Characteristics

(T_C = 25°C unless otherwise noted)

Characteristic	Symbol	Min.	Max.	Unit
Off Characteristics				
Collector-Emitter Sustaining Voltage (I_C = 100 mAdc, I_B = 0)	$V_{CEO(sus)}$	60	—	Vdc
Collector Cutoff Current (V_{CE} = 60 Vdc, I_B = 0) (V_{CE} = 100 Vdc, I_B = 0)	I_{CEO}	—	20	μA
Collector Cutoff Current (V_{CE} = 60 Vdc, $V_{BE(off)}$ = 1.5 Vdc) (V_{CE} = 60 Vdc, $V_{BE(off)}$ = 1.5 Vdc, T_C = 150°C)	I_{CEX}	— —	20 200	μA
Collector Cutoff Current (V_{CB} = 60 Vdc, I_E = 0)	I_{CBO}	—	20	μA
Emitter Cutoff Current (V_{BE} = 5.0 Vdc, I_C = 0)	I_{EBO}	—	2.0	mAdc

Contd...

On Characteristics				
DC Current Gain (I_C = 4.0 Adc, V_{CE} = 4.0 Vdc) (I_C = 8.0 Adc, V_{CE} = 4.0 Vdc)	h_{FE}	 1000 100	 20,000 —	—
Collector-Emitter Saturation Voltage (I_C = 4.0 Adc, I_B = 16 mAdc) (I_C = 8.0 Adc, I_B = 80 mAdc)	$V_{CE(sat)}$	 — —	 2.0 4.0	Vdc
Base-Emitter Saturation Voltage (I_C = 8.0 Adc, I_B = 80 mAdc)	$V_{BE(sat)}$	 —	 4.5	Vdc
Base-Emitter On Voltage (I_C = 4.0 Adc, V_{CE} = 4.0 Vdc)	$V_{BE(on)}$	 —	 2.8	Vdc
Dynamic Characteristics				
Small Signal Current Gain (I_C = 3.0 Adc, V_{CE} = 4.0 Vdc, f = 1.0 MHz)	$\|h_{fe}\|$	 4.0	 —	
Output Capacitance (V_{CB} = 10 Vdc, I_E = 0, f = 0.1 MHz)	C_{ob}	 —	 300	pF
Small-Signal Current Gain (I_C = 3.0 Adc, V_{CE} = 4.0 Vdc, f = 1.0 kHz)	h_{fe}	 300	 —	—

C.5 SL100 NPN SILICON MEDIUM POWER GENERAL PURPOSE TRANSISTOR

Maximum Ratings

Characteristic	Symbol	Min.	Max.	Unit
Collector-Emitter Voltage	BV_{CEO}	50	—	V
Collector-Base Voltage	BV_{CBO}	60	—	V
Emitter-Base Voltage	BV_{EBO}	5.0	—	V
Total Power Dissipation @ T_A = 25°C	P_D		800	mW
Collector Current	I_C		0.5	A
Operating and Storage Junction Temperature	T_j, T_{stg}		–65 to 200	°C

Electrical Characteristics

(T_A = 25°C unless otherwise noted)

Characteristic	Symbol	Min.	Max.	Unit
Off Characteristics				
Collector-Emitter Breakdown Voltage (I_C = 10 mAdc, I_B = 0)	BV_{CEO}	50	—	V
Collector-Base Breakdown Voltage (I_C = 100 μA, I_E = 0)	BV_{CBO}	60	—	V
Emitter-Base Breakdown Voltage (I_E = 100 μA, I_C = 0)	BV_{EBO}	5	—	V
Collector Cut-off Current (V_{CB} = 40 Vdc, I_E = 0)	I_{CBO}	—	50	nA
Emitter-Cutoff Current (V_{EB} = 4 Vdc, I_C = 0)	I_{EBO}	—	25	nA
On Characteristics				
DC Current Gain	h_{FE}			
(I_C = 10 mAdc, V_{CE} = 10 Vdc)		25	—	
(I_C = 150 mAdc, V_{CE} = 10 Vdc)		40	300	
Collector-Emitter Saturation Voltage (I_C = 150 mAdc, I_B = 15 mAdc)	$V_{CE(sat)}$	—	0.6	V
Base-Emitter Saturation Voltage (I_C = 150 mAdc, I_B = 15 mAdc)	$V_{BE(sat)}$	—	1.3	V
Small-Signal Characteristics				
Output Capacitance (V_{CB} = 10 V, I_E = 0, f = 140 kHz)	C_{obo}	—	20	pF

C.6 BD131 NPN 15W POWER TRANSISTOR

Features

- High current (max. 3 A)
- Low voltage (max. 45 V).

Description

NPN power transistor in a TO-126; SOT32 plastic package. PNP complement BD132.

Applications

- General purpose power applications

Limiting Values

In accordance with the Absolute Maximum Rating System (IEC 134)

Symbol	Parameter	Conditions	Min.	Max.	Unit
V_{CBO}	Collector-base voltage	Open emitter	—	70	V
V_{CEO}	Collector-emitter voltage	Open base	—	45	V
V_{EBO}	Emitter-base voltage	Open collector	—	6	V
I_C	Collector current (DC)		—	3	A
I_{CM}	Peak collector current		—	6	A
I_{BM}	Peak base current		—	0.5	A
P_{tot}	Total power dissipation	$T_{mb} \leq 60°C$	—	15	W
T_{stg}	Storage temperature		–65	+150	°C
T_j	Junction temperature		—	150	°C
T_{amb}	Operating ambient temperature		–65	+150	°C

Thermal Characteristics

Symbol	Parameter	Value	Unit
$R_{th\ j\text{-}a}$	Thermal resistance from junction to ambient	100	K/W
$R_{th\ j\text{-}mb}$	Thermal resistance from junction to mounting base	6	K/W

Characteristics

$T_j = 25°C$ unless otherwise specified.

Symbol	Parameter	Conditions	Min.	Max.	Unit
I_{CBO}	Collector cut-off current	$I_E = 0$; $V_{CB} = 50$ V	—	50	nA
		$I_E = 0$; $V_{CB} = 50$ V; $T_j = 150°C$	—	10	μA
I_{EBO}	Emitter cut-off current	$I_C = 0$; $V_{EB} = 5$ V	—	50	nA

Contd...

Symbol	Parameter	Conditions	Min.	Max.	Unit
h_{FE}	DC current gain	I_C = 0.5 A; V_{CE} = 12 V	40	—	
		I_C = 2 A; V_{CE} = 1 V	20	—	
V_{CEsat}	Collector-emitter saturation voltage	I_C = 0.5 A; I_B = 50 mA	—	300	mV
		I_C = 2A; I_B = 200 mA	—	700	mV
V_{BEsat}	Base-emitter saturation voltage	I_C = 0.5 A; I_B = 50 mA	—	1.2	V
		I_C = 2 A; I_B = 200 mA	—	1.5	V
f_T	Transition frequency	I_C = 0.25 A; V_{CE} = 5 V; f = 100 MHz; T_{amb} = 25°C	60	—	MHz

C.7 BD132 PNP 15W POWER TRANSISTOR

Features

- High current (max. 3 A)
- Low voltage (max. 45 V).

Applications

- General purpose power applications.

Description

PNP power transistor in a TO-126; SOT32 plastic package. NPN complement: BD131.

Quick Reference Data

Symbol	Parameter	Conditions	Min.	Max.	Unit
V_{CBO}	Collector-base voltage	Open emitter	—	–45	V
V_{CEO}	Collector-emitter voltage	Open base	—	–45	V
I_{CM}	Peak collector current		—	–6	A
P_{tot}	Total power dissipation	$T_{mb} \leq 60°C$	—	15	W
h_{FE}	DC current gain	$I_C = –0.5$ A; $V_{CE} = –12$ V	40	—	
f_T	Transition frequency	$I_C = –0.25$ A; $V_{CE} = –5$ V; f = 100 MHz	60	—	MHz

Limiting Values

In accordance with the Absolute Maximum Rating System (IEC 134).

Symbol	Parameter	Conditions	Min.	Max.	Unit
V_{CBO}	Collector-base voltage	Open emitter	—	–45	V
V_{CEO}	Collector-emitter voltage	Open base	—	–45	V
V_{EBO}	Emitter-base voltage	Open collector	—	–4	V
I_C	Collector current (DC)		—	–3	A
I_{CM}	Peak collector current		—	–6	A
I_{BM}	Peak base current		—	–0.5	A
P_{tot}	Total power dissipation	$T_{mb} \leq 60°C$	—	15	W
T_{stg}	Storage temperature		–65	+150	°C
T_j	Junction temperature		—	150	°C
T_{amb}	Operating ambient temperature		–65	+150	°C

Thermal Characteristics

Symbol	Parameter	Conditions	Value	Unit
$R_{th\ j\text{-}a}$	Thermal resistance from junction to ambient	note 1	100	K/W
$R_{th\ j\text{-}mb}$	Thermal resistance from junction to mounting base		6	K/W

Note 1: Refer to TO-126, SOT32 standard mounting conditions.

APPENDIX D: OP-AMPS AND LINEAR ICS DATASHEETS

D.1 LM741 OPERATIONAL AMPLIFIER

Features

- Overload protection on the input and output
- No latch-up when the common-mode range is exceeded

Applications

- Comparators
- DC Amplifiers
- Integrator or Differentiators
- Multivibrators
- Summing Amplifiers
- Active Filters

Description

The LM741 series are general-purpose operational amplifiers which feature improved performance over industry standards like the LM709. They are direct, plug-in replacements for the 709C, LM201, MC1439, and 748 in most applications.

The amplifiers offer many features which make their application nearly foolproof: overload protection on the input and output, no latch-up when the common-mode range is exceeded, as well as freedom from oscillations.

The LM741C is identical to the LM741 and LM741A except that the LM741C has their performance ensured over a 0°C to +70°C temperature range, instead of –55°C to +125°C.

Pin Configuration and Functions

LMC Package 8-Pin TO-99
Top View

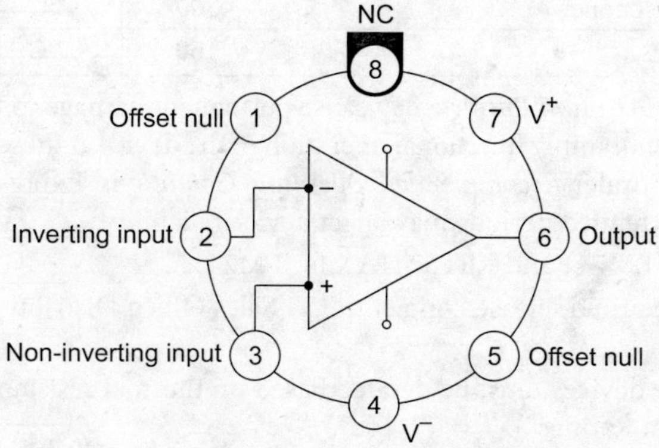

LM741H is available per JM385 10/10101

NAB Package
8-Pin CDIP or PDIP
Top View

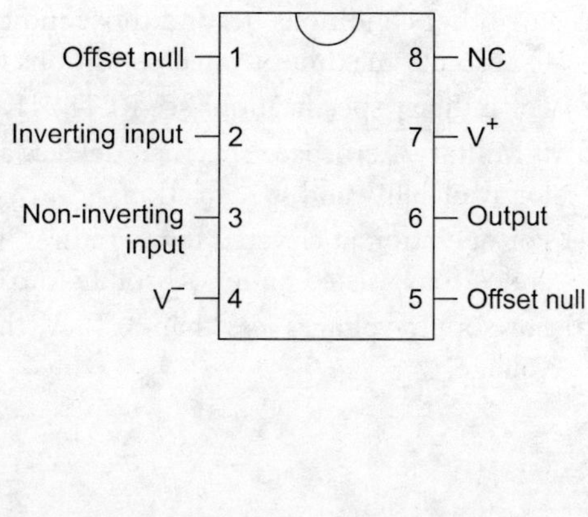

Pin Functions

Pin		I/O	Description
Name	No.		
Inverting input	2	I	Inverting signal input
NC	8	N/A	No connect, should be left floating
Non-inverting input	3	I	Non-inverting signal input
Offset null Offset null	1, 5	I	Offset nul pin used to eliminate the offset voltage and balance the input voltages
Output	6	O	Amplified signal output
V+	7	I	Positive supply voltage
V–	4	I	Negative supply voltage

Specifications

Absolute Maximum Ratings

Over operating free-air temperature range (unless otherwise noted)(1)(2)(3)

		Min	Max	Unit
Supply voltage	LM741, LM741A		±22	V
	LM741C		±18	
Power dissipation(4)			500	mW
Differential input voltage			±30	V
Input voltage(5)			±15	V
Output short circuit duration		Continuous		
Operating temperature	LM741, LM741A	–50	125	°C
	LM741C	0	70	
Junction temperature	LM741, LM741A		150	°C
	LM741C		100	
Soldering information	PDIP package (10 seconds)		260	°C
	CDIP or TO-99 package (10 seconds)		300	°C
Storage temperature, T_{stg}		–65	150	°C

(1) Stresses beyond those listed under *Absolute Maximum Ratings* may cause permanent damage to the device. These are stress ratings only, which do not imply functional operation of the device at these or any other conditions beyond those indicated under *Recommended Operating Conditions*. Exposure to absolute-maximum-rated conditions for extended periods may affect device reliability.

(2) For military specifications see RETS741X for LM741 and RETS741AX for LM741A.

(3) If Military/Aerospace specified devices are required, please contact the TI Sales Office/Distributors for availability and specifications.

(4) For operation at elevated temperatures, these devices must be derated based on thermal resistance, and T_j max. (listed under "Absolute Maximum Ratings"). $T_j = T_A + (\theta_{jA} P_D)$.

(5) For supply voltages less than ± 15 V, the absolute maximum input voltage is equal to the supply voltage.

ESP Ratings

			Value	Unit
$V_{(ESD)}$	Electrostatic discharge	Human body model (HBM), per ANSI/ESDA/JEDEC JS-001[1]	±400	V

(1) Level listed above is the passing level per ANSI, ESDA. and JEDEC JS-001. JEDEC document JEP155 states that 500-V HBM allows safe manufacturing with a standard ESD control process.

Recommended Operating Conditions

Over operating free-air temperature range (unless otherwise noted)

		Min	Nom	Max	Unit
Supply voltage (VDD-GND)	LM741, LM741A	±10	±15	±22	V
	LM741C	±10	±15	±18	
Temperature	LM741, LM741A	–55		125	°C
	LM741C	0		70	

Thermal Information

Thermal Metric[1]		LM741			Unit
		LMC(TO-99)	NAB(CDIP)	P(PDIP)	
		8 Pins	8 Pins	8 Pins	
$R_{\theta JA}$	Junction-to-ambient thermal resistance	170	100	100	°C/W
$R_{\theta JC(top)}$	Junction-to-case (top) thermal resistance	25	—	—	°C/W

Electrical Characteristics, LM741[1]

Parameter	Test Conditions		Min	Typ	Max	Unit
Input offset voltage	$R_S \leq 10$ kΩ	$T_A = 25°C$		1	5	mV
		$T_{Amin} \leq T_A \leq T_{Amax}$			6	mV
Input offset voltage adjustment range	$T_A = 25°C$, $V_S = \pm 20$ V			±15		mV
Input offset current	$T_A = 25°C$			20	200	nA
	$T_{Amin} \leq T_A \leq T_{Amax}$			85	500	
Input bias current	$T_A = 25°C$			80	500	nA
	$T_{Amin} \leq T_A \leq T_{Amax}$				1.5	μA
Input resistance	$T_A = 25°C$, $V_S = \pm 20$ V		0.3	2		mΩ
Input voltage range	$T_{Amin} \leq T_A \leq T_{Amax}$		±12	±13		V
Large signal voltage gain	$V_S = \pm 15$ V, $V_O = \pm 10$ V, $R_L \geq 2$ kΩ	$T_A = 25°C$	50	200		V/mV
		$T_{Amin} \leq T_A \leq T_{Amax}$	25			
Output voltage swing	$V_S = \pm 15$ V	$R_L \geq 10$ kΩ	±12	±14		V
		$R_L \geq 2$ kΩ	±10	±13		
Output short circuit current	$T_A = 25°C$			25		mA
Common-mode rejection ratio	$R_S \leq 10$ Ω, $V_{CM} = \pm 12$ V, $T_{Amin} \leq T_A \leq T_{Amax}$		80	95		dB
Supply voltage rejection ratio	$V_S = \pm 20$ V to $V_S = \pm 5$ V, $R_S \leq 10$ Ω, $T_{Amin} \leq T_A \leq T_{Amax}$		86	96		dB

Contd...

Parameter		Test Conditions		Min	Typ	Max	Unit
Transient response	Rise time	$T_A = 25°C$, unity gain			0.3		μs
	Overshoot				5%		
Slew rate		$T_A = 25°C$, unity gain			0.5		V/μs
Supply current		$T_A = 25°C$			1.7	2.8	mA
Power consumption		$V_S = \pm 15$ V	$T_A = 25°C$		50	85	mW
			$T_A = T_{Amin}$		60	100	
			$T_A = T_{Amax}$		45	75	

(1) Unless otherwise specified, these specifications apply for $V_S = \pm 15$ V, $-55°C \le T_A \le +125°C$ (LM741/LM741A). For the LM741C/LM741E, these specifications are limited to $0°C \le T_A \le +70°C$.

Electrical Characteristics, LM741A[(1)]

Parameter		Test Conditions		Min	Typ	Max	Unit
Input offset voltage		$R_S \le 10\ k\Omega$	$T_A = 25°C$		0.8	3	mV
			$T_{Amin} \le T_A \le T_{Amax}$			4	mV
Average input offset voltage drift						15	μV/°C
Input offset voltage adjustment range		$T_A = 25°C$, $V_S = \pm 20$ V		±10			mV
Input offset current		$T_A = 25°C$			3	30	nA
		$T_{Amin} \le T_A \le T_{Amax}$				70	
Average input offset current drift						0.5	nA/°C
Input bias current		$T_A = 25°C$			30	80	nA
		$T_{Amin} \le T_A \le T_{Amax}$				0.21	μA
Input resistance		$T_A = 25°C$, $V_S = \pm 20$ V		1	6		MΩ
		$T_{Amin} \le T_A \le T_{Amax}$, $V_S = \pm 20$ V		0.5			
Large signal voltage gain		$V_S = \pm 20$ V, $V_O = \pm 15$ V, $R_L \ge 2\ k\Omega$	$T_A = 25°C$	50			V/mV
			$T_{Amin} \le T_A \le T_{Amax}$	32			
		$V_S = \pm 5$ V, $V_O = \pm 2$ V, $R_L \ge 2\ k\Omega$, $T_{Amin} \le T_A \le T_{Amax}$		10			
Output voltage swing		$V_S = \pm 20$ V	$R_L \ge 10\ k\Omega$	±16			V
			$R_L \ge 2\ k\Omega$	±15			
Output short circuit current		$T_A = 25°C$		10	25	35	mA
		$T_{Amin} \le T_A \le T_{Amax}$		10		40	
Common-mode rejection ratio		$R_S \le 50\ \Omega$, $V_{CM} = \pm 12$ V, $T_{Amin} \le T_A \le T_{Amax}$		80	95		dB
Supply voltage rejection ratio		$V_S = \pm 20$ V to $V_S = \pm 5$ V, $R_S \le 50\ \Omega$, $T_{Amin} \le T_A \le T_{Amax}$		86	96		dB
Transient response	Rise time	$T_A = 25°C$, unity gain			0.25	0.8	μs
	Overshoot				6%	20%	
Bandwidth[(2)]		$T_A = 25°C$		0.437	1.5		MHz
Slew rate		$T_A = 25°C$, unity gain		0.3	0.7		V/μs

Contd...

Parameter	Test Conditions		Min	Typ	Max	Unit
Power consumption	$V_S = \pm 20$ V	$T_A = 25°C$		80	150	mW
		$T_A = T_{Amin}$			165	mW
		$T_A = T_{Amax}$			135	mW

(2) Calculated value from: BW (MHz) = 0.35/Rise Time (μs).

Electrical Characteristics, LM741C[(1)]

Parameter	Test Conditions		Min	Typ	Max	Unit
Input offset voltage	$R_S \le 10\ k\Omega$	$T_A = 25°C$		2	6	mV
		$T_{Amin} \le T_A \le T_{Amax}$			7.5	mV
Input offset voltage adjustment range	$T_A = 25°C$, $V_S = \pm 20$ V			±15		mV
Input offset current	$T_A = 25°C$			20	200	nA
	$T_{Amin} \le T_A \le T_{Amax}$				300	nA
Input bias current	$T_A = 25°C$			80	500	nA
	$T_{Amin} \le T_A \le T_{Amax}$				0.8	μA
Input resistance	$T_A = 25°C$, $V_S = \pm 20$ V		0.3	2		MΩ
Input voltage range	$T_A = 25°C$		±12	±13		V
Large signal voltage gain	$V_S = \pm 15$ V, $V_O = \pm 10$ V, $R_L \ge 2\ k\Omega$	$T_A = 25°C$	20	200		V/mV
		$T_{Amin} \le T_A \le T_{Amax}$	15			V/mV
Output voltage swing	$V_S = \pm 15$ V	$R_L \ge 10\ k\Omega$	±12	±14		V
		$R_L \ge 2\ k\Omega$	±10	±13		V
Output short circuit current	$T_A = 25°C$			25		mA
Common-mode rejection ratio	$R_S \le 10\ \Omega$, $V_{CM} = \pm 12$ V, $T_{Amin} \le T_A \le T_{Amax}$		70	90		dB
Supply voltage rejection ratio	$V_S = \pm 20$ V to $V_S = \pm 5$ V, $R_S \le 10\ \Omega$, $T_{Amin} \le T_A \le T_{Amax}$		77	96		dB
Transient response: Rise time	$T_A = 25°C$, unity gain			0.3		μs
Transient response: Overshoot	$T_A = 25°C$, unity gain			5%		
Slew rate	$T_A = 25°C$, unity gain			0.5		V/μs
Supply current	$T_A = 25°C$			1.7	2.8	mA
Power consumption	$V_S = \pm 15$ V, $T_A = 25°C$			50	85	mW

(1) Unless otherwise specified, these specifications apply for $V_S = \pm 15$ V, $-55°C \le T_A \le +125°C$ (LM741/LM741A). For the LM741C/LM741E, these specifications are limited to $0°C \le T_A \le +70°C$.

D.2 LM118-N/LM218-N/LM318-N OPERATIONAL AMPLIFIERS

Description

The LM118 series are precision high speed operational amplifiers designed for applications requiring wide bandwidth and high slew rate. They feature a factor of ten increase in speed over general purpose devices without sacrificing DC performance.

The LM118 series has internal unity gain frequency compensation. This considerably simplifies its application since no external components are necessary for operation. However, unlike most internally compensated amplifiers, external frequency compensation may be added for optimum performance. For inverting applications, feed forward compensation will boost the slew rate to over 150V/μs and almost double the bandwidth. Overcompensation can be used with the amplifier for greater stability when maximum bandwidth is not needed. Further, a single capacitor can be added to reduce the 0.1% settling time to under 1 μs.

The high speed and fast settling time of these op amps make them useful in A/D converters, oscillators, active filters, sample and hold circuits, or general purpose amplifiers. These devices are easy to apply and offer an order of magnitude better AC performance than industry standards such as the LM709.

The LM218-N is identical to the LM118 except that the LM2I8-N has its performance specified over a –25°C to +85°C temperature range. The LM3I8-N is specified from 0°C to +70°C

Features

- 15 MHz Small Signal Bandwidth
- Ensured 50 V/μs Slew Rate
- Maximum Bias Current of 250 nA
- Operates from Supplies of ± 5 V to ± 20 V
- Internal Frequency Compensation
- Input and Output Overload Protected
- Pin Compatible with General Purpose Op Amps

Pin Diagram

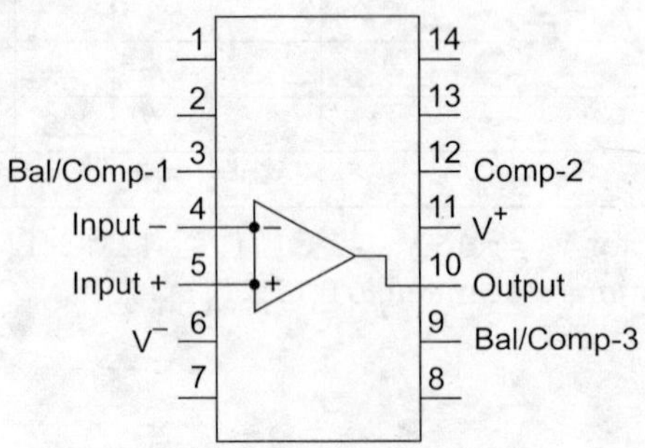

Available per JM38510/10107

Dual-In-Line Package
(Top View)

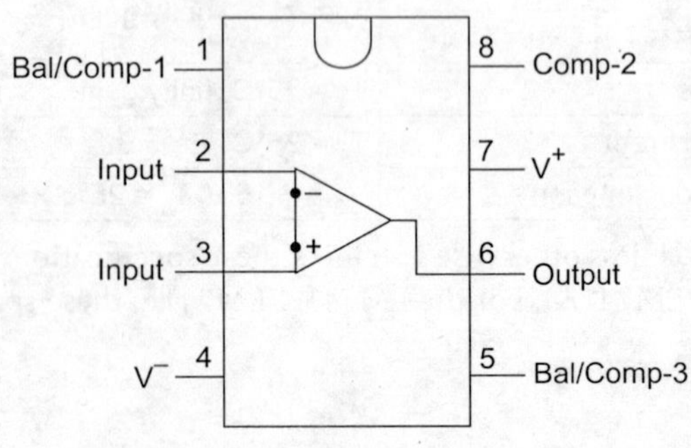

Available per JM38510/10107

Dual-In-Line Package
(Top View)

Contd...

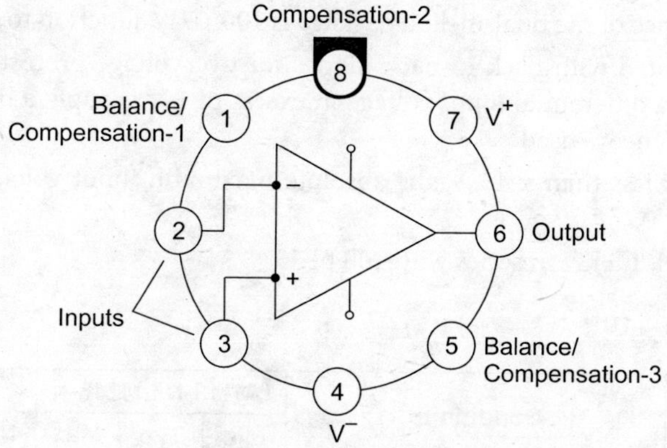

Pin connections shown on shematic diagram and typical applications are for TO-99 package

TO-99 Package
(Top View)

Absolute Maximum Ratings[(1)(2)]

Supply voltage	± 20 V
Power dissipation[(3)]	500 mW
Differential input current[(4)]	± 10 mA
Input voltage[(5)]	± 15 V
Output short-circuit duration	Continuous
Operating temperature range	
LM118-N	– 55°C to + 125°C
LM218-N	– 25°C to + 85°C
LM318-N	0°C to + 70°C
Storage temperature range	– 65°C to + 150°C
Lead temperature (Soldering, 10 sec.)	
TO-99 package	300°C
PDIP package	260°C
Soldering information	
Dual-in-line package	
Soldering (10 sec.)	260°C
SOIC package	
Vapour phase (60 sec.)	215°C
Infrared (15 sec.)	220°C
ESD tolerance[(6)]	2000 V

(1) Refer to RETS 118X for LM118H and LM118J military specifications

(2) If Military/Aerospace specified devices are required, please contact the TI Sales Office/Distributors for availability and specifications.

(3) The maximum junction temperature of the LM118-N is 150°C, the LM218-N is 110°C, and the LM318-N is 110°C. For operating at elevated temperatures, devices in the LMC package must be

derated based on a thermal resistance of 160°C/W, junction to ambient, or 20°C/W, junction to case. The thermal resistance of the dual-in-line package is 100°C/W, junction to ambient.

(4) The inputs are shunted with back-to-back diodes for overvoltage protection. Therefore, excessive current will flow if a differential input voltage in excess of 1 V is applied between the inputs unless some limiting resistance is used.

(5) For supply voltages less than ± 15 V, the absolute maximum input voltage is equal to the supply voltage.

(6) Human body model, 1.5 kΩ in series with 100 pF.

Electrical Characteristics[1]

Parameter	Conditions	LM118-N/LM218-N			LM318-N			Units
		Min	Typ	Max	Min	Typ	Max	
Input Offset voltage	T_A = 25°C		2	4		4	10	mV
Input Offset Current	T_A = 25°C		6	50		30	200	nA
Input Bias Current	T_A = 25°C		120	250		150	500	nA
Input Resistance	T_A = 25°C	1	3		0.5	3		MΩ
Supply Current	T_A = 25°C		5	8		5	10	mA
Large Signal Voltage Gain	T_A = 25°C, V_S = ± 15 V V_{OUT} = ± 10 V, $R_L \geq 2$ kΩ	50	200		25	200		V/mV
Slew Rate	T_A = 25°C, V_S = ± 15 V, A_V = 1[2]	50	70		50	70		V/µs
Small Signal Bandwidth	T_A = 25°C, V_S = ± 15 V		15			15		MHz
Input Offset Voltage				6			15	mV
Input Offset Current				100			300	nA

(1) These specifications apply for $\pm 5\text{ V} \leq V_S \leq \pm 20\text{ V}$ and $-55°\text{C} \leq T_A \leq +125°\text{C}$ (LM118-N), $-25°\text{C} < T_A < +85°\text{C}$ (LM218-N), and $0°\text{C} \leq T_A \leq +70°\text{C}$ (LM318-N). Also, power supplies must be bypassed with 0.1 pF disc capacitors.

(2) Slew rate is tested with $V_S = \pm 15$ V. The LM118-N is in a unity-gain non-inverting configuration. V_{IN} is stepped from – 7.5 V to + 7.5 V and vice versa. The slew rates between – 5.0 V and + 5.0 V and vice versa are tested and specified to exceed 50 V/µs.

Electrical Characteristics[1] (Continued)

Parameter	Conditions	LM118-N/LM218-N			LM318-N			Units
		Min	Typ	Max	Min	Typ	Max	
Input Bias Current				500			750	nA
Supply Current	T_A = 125°C		4.5	7				mA
Large Signal Voltage Gain	V_S = ± 15 V, V_{OUT} = ± 10 V, $R_L \geq 2$ kΩ	25			20			V/mV
Output Voltage Range	V_S = ± 15 V, R_L = 2 kΩ	± 12	± 13		± 12	± 13		V
Input Voltage Range	V_S = ± 15 V	± 11.5			± 11.5			V
Common-Mode Rejection Ratio		80	100		70	100		dB
Supply Voltage Rejection Ratio		70	80		65	80		dB

D.3 LM555 TIMER

Features

- Direct Replacement for SE555/NE555
- Timing from Microseconds through Hours
- Operates in Both Astable and Monostable Modes
- Adjustable Duty Cycle
- Output Can Source or Sink 200 mA
- Output and Supply TTL Compatible
- Temperature Stability Better than 0.005% per °C
- Normally On and Normally Off Output
- Available in 8-pin VSSOP Package

Applications

- Precision Timing
- Pulse Generation
- Sequential Timing
- Time Delay Generation
- Pulse Width Modulation
- Pulse Position Modulation
- Linear Ramp Generator

Pin Configuration and Functions

D, P and DGK Packages
8-Pin PDIP, SOIC and VSSOP
Top View

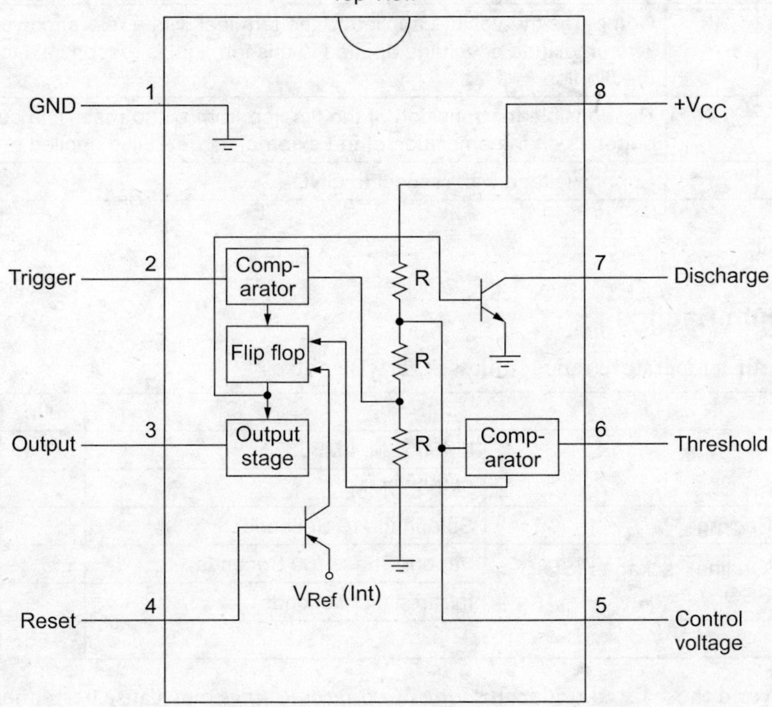

Description

The LM555 is a highly stable device for generating accurate time delays or oscillation. Additional terminals are provided for triggering or resetting if desired. In the time delay mode of operation, the time is precisely controlled by one external resistor and capacitor. For a stable operation as an oscillator, the free running frequency and duty cycle are accurately controlled with two external resistors and one capacitor. The circuit may be triggered and reset on falling waveforms, and the output circuit can source or sink up to 200 mA or drive TTL circuits.

Pin Functions

Pin		I/O	Description
No.	Name		
5	Control Voltage	I	Controls the threshold and trigger levels. It determines the pulse width of the output waveform. An external voltage applied to this pin can also be used to modulate the output waveform
7	Discharge	I	Open collector output which discharges a capacitor between intervals (in phase with output). It toggles the output from high to low wren voltage reaches 2/3 of the supply voltage
1	GND	O	Ground reference voltage
3	Output	O	Output driven waveform
4	Reset	I	Negative pulse applied to this pin to disable or reset the timer. When not used for reset purposes, it should be connected to VCC to avoid false triggering
6	Threshold	I	Compares the voltage applied to the terminal with a reference voltage of 2/3 Vcc. The amplitude of voltage applied to this terminal is responsible tor the set state of the flip-flop
2	Trigger	I	Responsible for transition of the flip-flop from set to reset The output of the timer depends on the amplitude of the external trigger pulse applied to this pin
8	V^+	I	Supply voltage with respect to GND

Specifications

Absolute Maximum Ratings

Over operating free-air temperature range (unless otherwise noted)[(1)(2)]

			Min	Max	Unit
Power Dissipation[(3)]		LM555CM, LM555CN[(4)]		1180	mW
		LM555CMM		613	mW
Soldering Information	PDIP Package	Soldering (10 Seconds)		260	°C
	Small Outline Packages (SOIC and VSSOP)	Vapour Phase (60 Seconds)		215	°C
		Infrared (15 Seconds)		220	°C
			–65	150	°C

(1) Stresses beyond those listed under *Absolute Maximum Ratings* may cause permanent damage to the device. These are stress ratings only, which do not imply functional operation of the device at these or

any other conditions beyond those indicated under Recommended Operating Conditions- Exposure to absolute-maximum-rated conditions for extended periods may affect device reliability.

(2) If Military/Aerospace specified devices are required, please contact the TI Sales Office/Distributors for availability and specifications

(3) For operating at elevated temperatures the device must be derated above 25°C based on a 150°C maximum junction temperature and a thermal resistance of 106°C/W (PDIP), 170°C/W (SOIC-8), and 204°C/W (VSSOP) junction to ambient.

(4) Refer to RETS555X drawing of military LM555H and LM555J versions for specifications.

ESD Ratings

			Value	Unit
$V_{(ESD)}$	Electrostatic discharge	Human-body model (HBM), per ANSI/ESDA/JEDEC JS-001(1)	± 500(2)	V

(1) JEDEC document JEP155 states that 500-V HBM allows safe manufacturing with a standard ESD control process

(2) The ESD information listed is for the SOIC package.

Recommended Operating Conditions

Over operating free-air temperature range (unless otherwise noted)

	Min	Max	Unit
Supply voltage		18	V
Temperature, T_A	0	70	°C
Operating junction temperature, T_J		70	°C

Thermal Information

Thermal Metric(1)		LM555			Unit
		PDIP	SOIC	VSSOP	
		8 Pins			
$R_{\theta JA}$	Junction-to-ambient thermal resistance	106	170	204	°C/W

Electrical Characteristics

(T_A = 25°C, V_{cc} = 5 V to 15 V, unless otherwise specified)(1)(2)

Parameter	Test Conditions	Min	Typ	Max	Unit
Supply Voltage		4.5		16	V
Supply Current	V_{CC} = 5 V, $R_L = \infty$		3	6	mA
	V_{CC} = 15 V, $R_L = \infty$ (Low State)		10	15	
Timing Error, Monostable					
Initial Accuracy			1%		
Drift with Temperature	R_A = 1 k to 100 kΩ		50		ppm/°C
	C = 0.1 μF, (4)				

Contd...

Parameter	Test Conditions	Min	Typ	Max	Unit
Accuracy over Temperature			1.5%		
Drift with Supply			0.1%		V
Timing Error, Astable					
Initial Accuracy			2.25		
Drift with Temperature	R_A, R_B = 1 k to 100 kΩ		150		ppm/°C
	C = 0.1 μF, (4)				
Accuracy over Temperature			3.0%		
Drift with Supply			0.30%		/V
Threshold Voltage			0.667		× V_{CC}
Trigger Voltage	V_{CC} = 15 V		5		V
	V_{CC} = 5 V		1.67		
Trigger Current			0.5	0.9	μA
Reset Voltage		0.4	0.5	1	V
Reset Current			0.1	0.4	mA
Threshold Current	(5)		0.1	0.25	μA
Control Voltage Level	V_{CC} = 15 V	9	10	11	V
	V_{CC} = 5 V	2.6	3.33	4	
Pin 7 Leakage Output High			1	100	nA
Pin 7 Sat(6)					
Output Low	V_{CC} = 15 V, I_7 = 15 mA		180		mV
Output Low	V_{CC} = 4.5 V, I_7 = 4.5 mA		80	200	mV
Output Voltage Drop (Low)	V_{CC} = 15 V				
	I_{SINK} = 10 mA		0.1	0.25	V
	I_{SINK} = 50 mA		0.4	0.75	V
	I_{SINK} = 100 mA		2	2.5	V
	I_{SINK} = 200 mA		2.5		V
	V_{CC} = 5 V				
	I_{SINK} = 8 mA				V
	I_{SINK} = 5 mA		0.25	0.35	V
Output Voltage Drop (High)	I_{SOURCE} = 200 mA, V_{CC} = 15 V		12.5		V
	I_{SOURCE} = 100 mA, V_{CC} = 15 V	12.75	13.3		V
	V_{CC} = 5 V	2.75	3.3		V
Rise Time of Output			100		ns
Fall Time of Output			100		ns

(1) All voltages are measured with respect to the ground pin, unless otherwise specified

(2) *Absolute Maximum Ratings* indicate limits beyond which damage to the device may occur. Recommended *Operating Conditions* indicate conditions for which the device is functional, but do not ensure specific performance limits. *Electrical Characteristics* state DC and AC electrical specifications under particular test conditions which ensures specific performance limits. This

assumes that the device is within the *Recommended Operating Conditions.* Specifications are not ensured for parameters where no limit is given, however, the typical value is a good indication of device performance.

(3) Supply current when output high typically 1 mA less at $V_{CC} = 5$ V.

(4) Tested at $V_{CC} = 5$ V and $V_{CC} = 15$ V.

(5) This will determine the maximum value of $R_A + R_B$ for 15 V operation. The maximum total $(R_A + R_B)$ is 20 MΩ.

(6) No protection against excessive pin 7 current is necessary providing the package dissipation rating will not be exceeded.

D.4 LM7805/7815 3-TERMINAL 1A POSITIVE VOLTAGE REGULATORS

General Description

The LM78XX series of three terminal regulators is available with several fixed output voltages making them useful in a wide range of applications. One of these is local on card regulation, eliminating the distribution problems associated with single point regulation. The voltages available allow these regulators to be used in logic systems, instrumentation, HiFi, and other solid state electronic equipment. Although designed primarily as fixed voltage regulators these devices can be used with external components to obtain adjustable voltages and currents.

The LM78XX series is available in an aluminum TO-3 package which will allow over 1.0 A load current if adequate heat sinking is provided. Current limiting is included to limit the peak output current to a safe value. Safe area protection for the output transistor is provided to limit internal power dissipation. If internal power dissipation becomes too high for the heat sinking provided, the thermal shutdown circuit takes over preventing the IC from overheating.

Considerable effort was expanded to make the LM78XX series of regulators easy to use and minimize the number of external components. It is not necessary to bypass the output, although this does improve transient response. Input bypassing is needed only if the regulator is located far from the filter capacitor of the power supply.

Features

- Output current in excess of 1 A
- No external components required
- Internal short circuit current limit
- Internal thermal overload protection
- Output transistor safe area protection
- Available in the aluminum TO-3 package

Voltage Range

- LM7805C 5 V
- LM7815C 15 V

Connection Diagrams

Metal Can Package
TO-3 (K)
Aluminium

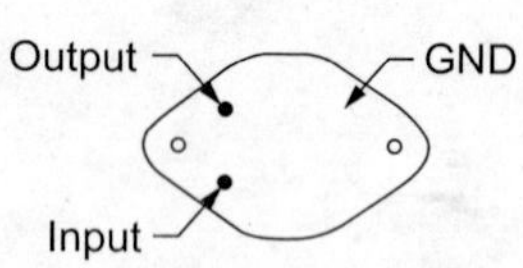

Bottom View

Plastic Package
TO-220-T

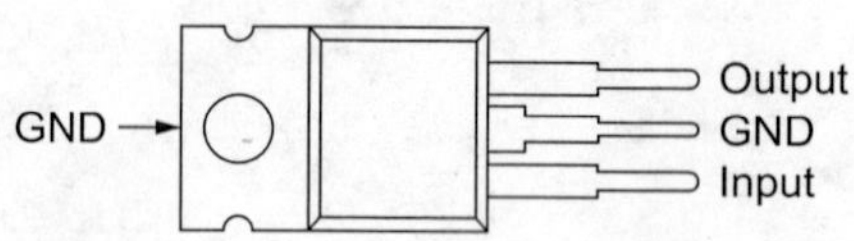

Top View

Fixed Output Voltage Regulator

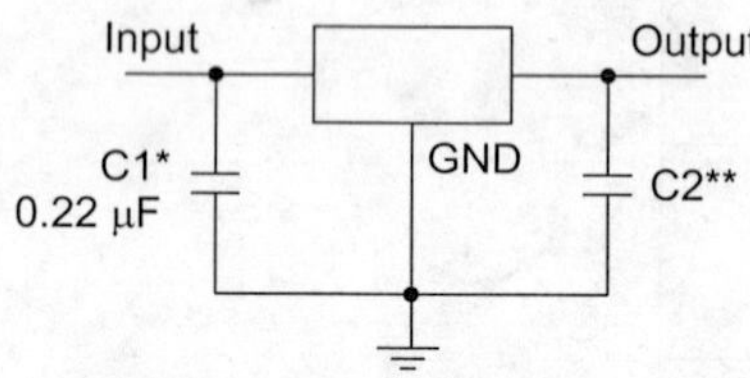

* Required if the regulator is located far from the power supply filter.

** Although no output capacitor is needed for stability, it does help transient response. (If needed, use 0.1 µF, ceramic disc).

Absolute Maximum Ratings

Input Voltage	
(V_O = 5 V and 15 V)	35 V
Internal Power Dissipation	Internally Limited
Operating Temperature Range (T_A)	0°C to +70°C
Maximum Junction Temperature	
(K Package)	150°C
(T Package)	150°C
Storage Temperature Range	– 65°C to + 150°C
Lead Temperature (Soldering, 10 sec.)	
TO-3 Package K	300°C
TO-220 Package T	230°C

Electrical Characteristics LM78XXC

0°C ≤ T_J ≤ 125°C unless otherwise noted.

Output Voltage				5 V			15 V			Units
Input Voltage (unless otherwise noted)				10 V			23 V			
Symbol	Parameter	Conditions		Min	Typ	Max	Min	Typ	Max	
V_O	Output Voltage	T_j = 25°C, 5 mA ≤ I_O ≤ 1 A		4.8	5	5.2	14.4	15	15.6	V
		P_D ≤ 15 W, 5 mA ≤ I_O ≤ 1 A		4.75		5.25	14.25		15.75	V
		V_{MIN} ≤ V_{IN} ≤ V_{MAX}		(7.5 ≤ V_{IN} ≤ 20)			(17.5 ≤ V_{IN} ≤ 30)			V
ΔV_O	Line Regulation	I_O = 500 mA	T_j = 25°C		3	50		4	150	mV
			ΔV_{IN}	(7 ≤ V_{IN} ≤ 25)			(17.5 ≤ V_{IN} ≤ 30)			V
			0°C ≤ T_j ≤ +125°C			50			150	mV
			ΔV_{IN}	(8 ≤ V_{IN} ≤ 20)			(18.5 ≤ V_{IN} ≤ 30)			V
		I_O ≤ 1A	T_j = 25°C			50			150	mV
			ΔV_{IN}	(7.5 ≤ V_{IN} ≤ 20)			(17.7 ≤ V_{IN} ≤ 30)			V
			0°C ≤ T_j ≤ +125°C			25			75	mV
			ΔV_{IN}	(8 ≤ V_{IN} ≤ 12)			(20 ≤ V_{IN} ≤ 26)			V
ΔV_O	Load Regulation	T_j = 25°C	5 mA ≤ I_O ≤ 1.5 A		10	50		12	150	mV
			250 mA ≤ I_O ≤ 750 mA			25			75	mV
		5 mA ≤ I_O ≤ 1 A, 0°C ≤ T_j ≤ +125°C				50			150	mV
I_Q	Quiescent Current	I_O ≤ 1 A	T_j = 25°C			8			8	mA
			0°C ≤ T_j ≤ + 125°C			8.5			8.5	mA
ΔI_Q	Quiescent Current Change	5 mA ≤ I_O ≤ 1 A				0.5			0.5	mA
		T_j = 25°C, I_O ≤ 1 A				1.0			1.0	mA
		V_{MIN} ≤ V_{IN} ≤ V_{MAX}		(7.5 ≤ V_{IN} ≤ 20)			(17.9 ≤ V_{IN} ≤ 20)			V
		I_O ≤ 500 mA, 0°C ≤ T_j ≤ + 125°C				1.0			1.0	mA
		V_{MIN} ≤ V_{IN} ≤ V_{MAX}		(7 ≤ V_{IN} ≤ 25)			(17.5 ≤ V_{IN} ≤ 30)			V

Contd...

Output Voltage				5 V			15 V			Units
Input Voltage (unless otherwise noted)				10 V			23 V			
Symbol	Parameter	Conditions		Min	Typ	Max	Min	Typ	Max	
V_N	Output Noise Voltage	$T_A = 25°C$, $10 \text{ Hz} \leq f \leq 100 \text{ kHz}$			40			90		μV
$\frac{\Delta V_{IN}}{\Delta V_{OUT}}$	Ripple Rejection	f = 120 Hz	$I_O \leq 1$ A, $T_j = 25°C$ or	62	80		54	70		dB
			$I_O \leq 500$ mA $0°C \leq T_j \leq +125°C$	62			54			dB
		$V_{MIN} \leq V_{IN} \leq V_{MAX}$		$(8 \leq V_{IN} \leq 18)$			$(18.5 \leq V_{IN} \leq 28.5)$			V
R_O	Dropout Voltage Output Resistance	$T_j = 25°C$, $I_{OUT} = 1$ A f = 1 kHz			2.0 8			2.0 19		V mΩ
	Short-Circuit Current	$T_j = 25°C$			2.1			1.2		A
	Peak Output Current	$T_j = 25°C$			2.4			2.4		A
	Average TC of V_{OUT}	$0°C \leq T_j \leq +125°C$, $I_O = 5$ mA			0.6			1.8		mV/°C
V_{IN}	Input Voltage Required to Maintain Line Regulation	$T_j = 25°C$, $I_O \leq 1$ A			7.5			17.7		V

D.5 LM7915 3-TERMINAL 1A NEGATIVE VOLTAGE REGULATOR

General Description

The LM79XX series of 3-terminal regulators is available with fixed output voltages of – 5 V, – 8 V, – 12 V, and – 15 V. These devices need only one external component—a compensation capacitor at the output. The LM79XX series is packaged in the TO-220 power package and is capable of supplying 1.5 A of output current.

These regulators employ internal current limiting safe area protection and thermal shutdown for protection against virtually all overload conditions.

Low ground pin current of the LM79XX series allows output voltage to be easily boosted above the preset value with a resistor divider. The low quiescent current drain of these devices with a specified maximum change with line and load ensures good regulation in the voltage boosted mode.

Features

- Thermal, short circuit and safe area protection
- High ripple rejection
- 4% tolerance on preset output voltage
- 1.5 A output current

Connection Diagrams

TO-220 Package

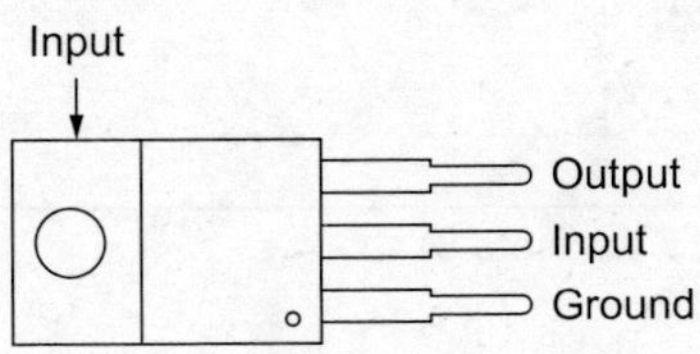

Front View

Typical Applications

Fixed Regulator

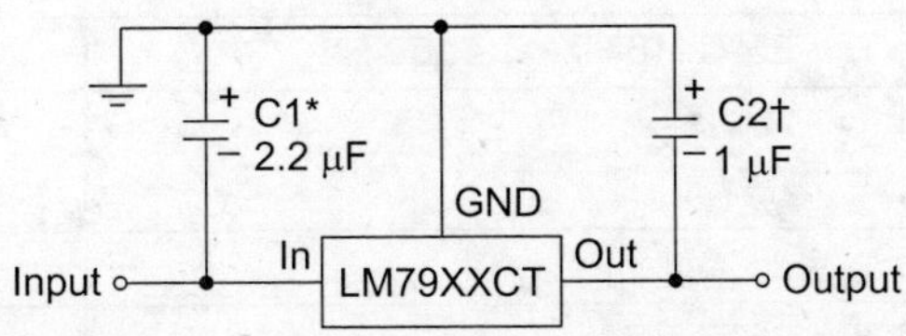

*Required if regulator is separated from filter capacitor by more than 3″. For value given, capacitor must be solid tantalum. 25 μF aluminum electrolytic may be substituted.

†Required for stability. For value given, capacitor must be solid tantalum. 25 μF aluminum electrolytic may be substituted. Values given may be increased without limit. For output capacitance in excess of 100 μF, a high current diode from input to output (1N4001, etc.) will protect the regulator from momentary input shorts.

Absolute Maximum Ratings

Input Voltage (V_O = – 15 V)	– 35 V
Input-Output Differential (V_O = – 15 V)	30 V
Power Dissipation	Internally Limited
Operating Junction Temperature Range	0°C to + 125°C
Storage Temperature Range	– 65°C to + 150°C
Lead Temperature (Soldering, 10 sec.)	230°C

Part Number			LM7915C			Units
Output Voltage			– 15 V			
Input Voltage (unless otherwise specified)			– 23 V			
Symbol	**Parameter**	**Conditions**	**Min**	**Typ**	**Max**	
V_O	Output Voltage	T_J = 25°C	–14.4	–15.0	–15.6	V
		5 mA ≤ I_{OUT} ≤ 1 A,	–14.25		–15.75	V
		P ≤ 15 W	(–30 ≤ V_{IN} ≤ –17.5)			V
ΔV_O	Line Regulation	T_J = 25°C, (Note 3)		5	100	mV
			(–30 ≤ V_{IN} ≤ –17.5)			V
				3	50	mV
			(–26 ≤ V_{IN} ≤ –20)			V
ΔV_O	Load Regulation	T_J = 25°C, (Note 3)				
		5 mA ≤ I_{OUT} ≤ 1.5 A		15	200	mV
		250 mA ≤ I_{OUT} ≤ 750 mA		5	75	mV
I_Q	Quiescent Current	T_J = 25°C		1.5	3	mA
ΔI_Q	Quiescent Current Change	With Line			0.5	mA
			(–30 ≤ V_{IN} ≤ –17.5)			V
		With Load, 5 mA ≤ I_{OUT} ≤ 1 A			0.5	mA
V_n	Output Noise Voltage	T_A = 25°C, 10 Hz ≤ f ≤ 100 Hz		375		μV
	Ripple Rejection	f = 120 Hz	54	70		dB
			(–30 ≤ V_{IN} ≤ –17.5)			V
	Dropout Voltage	T_J = 25°C, I_{OUT} = 1 A		1.1		V
I_{OMAX}	Peak Output Current	T_J = 25°C		2.2		A
	Average Temperature Coefficient of Output Voltage	I_{OUT} = 5 mA, 0°C ≤ T_J ≤ 100°C		–1.0		mV/°C

D.6 LM723 VOLTAGE REGULATOR

FEATURES

- 150 mA Output Current Without External Pass Transistor
- Output Currents in Excess of 10 A Possible by Adding External Transistors
- Input Voltage 40 V Max
- Output Voltage Adjustable from 2 V to 37 V
- Can be Used as Either a Linear or a Switching Regulator

Description

The LM723/LM723C is a voltage regulator designed primarily for series regulator applications. By itself, it will supply output currents up to 150 mA; but external transistors can be added to provide any desired load current. The circuit features extremely low standby current drain, and provision is made for either linear or foldback current limiting.

The LM723/LM723C is also useful in a wide range of other applications such as a shunt regulator, a current regulator or a temperature controller.

The LM723C is identical to the LM723 except that the LM723C has its performance ensured over a 0°C to +70°C temperature range, instead of −55°C to +125°C.

Connection Diagram

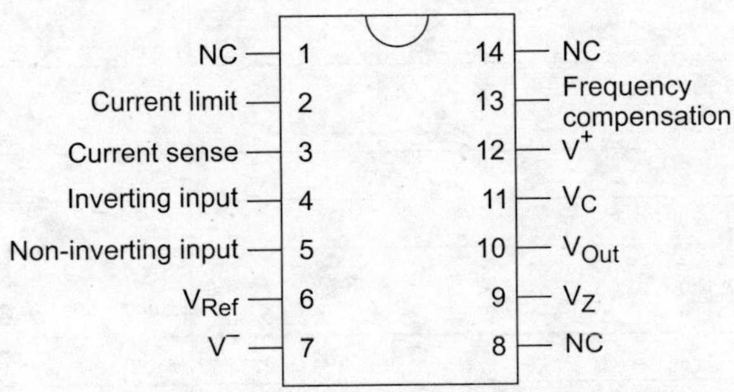

Figure 1. Top View CDIP Package or PDIP Package

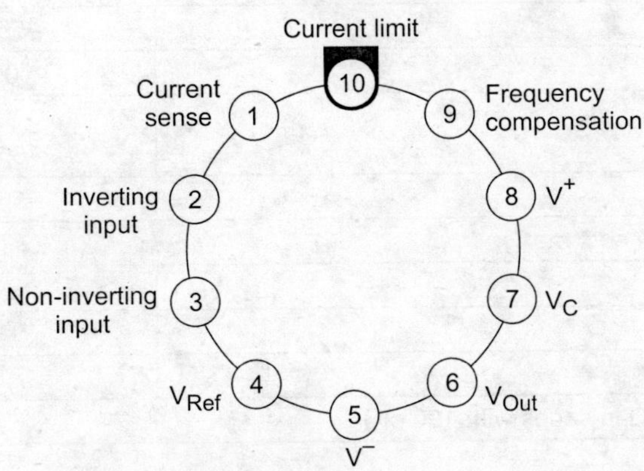

Note: Pin 5 connected to case.

Figure 2. Top View TO-100

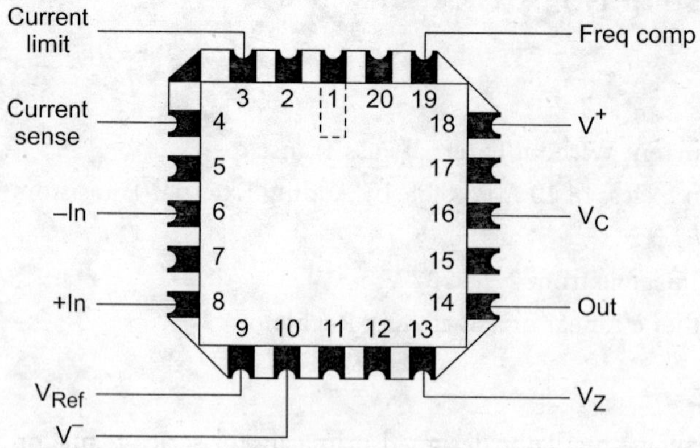

Figure 3. Top View

See Package NAJ0020A

Absolute Maximum Ratings

Pulse Voltage from V^+ to V^- (50 ms)	50 V
Continuous Voltage from V^+ to V^-	40 V
Input-Output Voltage Differential	40 V
Maximum Amplifier Input Voltage (Either Input)	8.5 V
Maximum Amplifier Input Voltage (Differential)	5 V
Current from V_Z	25 mA
Current from V_{REF}	15 mA
Internal Power Dissipation Metal Can[(3)]	800 mW
CDIP[(3)]	900 mW
PDIP[(3)]	660 mW
Operating Temperature Range	
LM723	–55°C to +150°C
LM723C	0°C to +70°C
Storage Temperature Range Metal Can	–65°C to +150°C
PDIP	–55°C to +150°C
Lead Temperature (Soldering, 4 sec. max.)	
Hermetic Package	300°C
Plastic Package	260°C
ESD Tolerance	1200 V
(Human body model, 15 kΩ in series with 100 pF)	

Electrical Characteristics[(1)(2)(3)(4)]

Parameter	Conditions	LM723			LM723C			Units
		Min	Typ	Max	Min	Typ	Max	
Line Regulation	V_{IN} = 12 V to V_{IN} = 15 V		0.01	0.1		0.01	0.1	$\%V_{OUT}$
	–55°C ≤ T_A ≤ +125°C			0.3				$\%V_{OUT}$
	0°C ≤ T_A ≤ +70°C						0.3	$\%V_{OUT}$
	V_{IN} = 12 V to V_{IN} = 40 V		0.02	0.2		0.1	0.5	$\%V_{OUT}$
Load Regulation	I_L = 1 mA to I_L = 50 mA		0.03	0.15		0.03	0.2	$\%V_{OUT}$
	–55°C ≤ T_A ≤ +125°C			0.6				$\%V_{OUT}$
	0°C ≤ T_A ≤ +70°C						0.6	$\%V_{OUT}$
Ripple Rejection	f = 50 Hz to 10 kHz, C_{REF} = 0		74			74		dB
	f = 50 Hz to 10 kHz, C_{REF} = 5 μF		86			86		dB

(1) Unless otherwise specified, T_A = 25°C, $V_{IN} = V^+ = V_C$ = 12 V, V^- = 0, V_{OUT} = 5 V, I_L = 1 mA, R_{SC} = 0, C_1 = 100 pF. C_{REF} = 0 and divider impedance as seen by error amplifier ≤ 10 kΩ connected. Line and load regulation specifications are given for the condition of constant chip temperature. Temperature drifts must be taken into account separately for high dissipation conditions.

(2) A military RETS specification is available on request. At the time of printing, the LM723 RETS specification complied with the Min and Max limits in this table. The LM723E, H, and J may also be procured as a Standard Military Drawing.

(3) Specified by correlation to other tests.

(4) L_1 is 40 turns of No. 20 enameled copper wire wound on Ferroxcube P36/22-3B7 pot core or equivalent with 0.009 in. air gap.

Electrical Characteristics

Parameter	Conditions	LM723			LM723C			Units
		Min	Typ	Max	Min	Typ	Max	
Average Temperature Coefficient of Output Voltage[(5)]	–55°C ≤ T_A ≤ +125°C		0.002	0.015				%/°C
	0°C ≤ T_A ≤ +70°C					0.003	0.015	%/°C
Short Circuit Current Limit	R_{SC} = 10 Ω, V_{OUT} = 0		65			65		mA
Reference Voltage		6.95	7.15	7.35	6.80	7.15	7.50	V
Output Noise Voltage	BW = 100 Hz to 10 kHz, C_{REF} = 0		86			86		μV/rms
	BW = 100 Hz to 10 kHz, C_{REF} = 5 μF		2.5			2.5		μV/rms
Long Term Stability			0.05			0.05		%/1000 hrs
Standby Current Drain	I_L = 0, V_{IN} = 30 V		1.7	3.5		17	4.0	mA
Input Voltage Range		9.5		40	9.5		40	V
Output Voltage Range		2.0		37	2.0		37	V
Input-Output Voltage Differential		3.0		38	3.0		38	V

Contd...

Parameter	Conditions	LM723			LM723C			Units
		Min	Typ	Max	Min	Typ	Max	
θ_{JA}	PDIP					105		°C/W
θ_{JA}	CDIP		150					°C/W
θ_{JA}	H10C Board Mount in Still Air		165			165		°C/W
θ_{JA}	H10C Board Mount in 400 LF/Min Air Flow		66			66		°C/W
θ_{JC}			22			22		°C/W

(5) For metal can applications where V_Z is required, an external 6.2 V zener diode should be connected in series with V_{OUT}

Index

Bibliography

- Amos Stanley William, *Principles of Transistor Circuits: Introduction to the Design of Amplifiers, Receivers, and Other Circuits*, Iliffe, 1959.
- Bar-Lev Adir, *Semiconductors and Electronic Devices*, Pearson Education, 1984.
- Boylestad Robert L. and Nashelsky Louis, *Electronic Devices and Circuit Theory*, Pearson Education Limited, 2013.
- Carusone Tony Chan, Johns David, and Martin Kenneth, *Analog Integrated Circuit Design*, John Wiley & Sons Inc., 2011.
- Cathey Jimmie J., *Theory and Problems of Electronic Devices and Circuits*, Schaum's Outline Series, Mcgraw-Hill, 2002.
- Clayton G.B., *Operational Amplifiers*, Newnes Butterworths, 1979.
- Comer David J. and Comer Donald T., *Fundamentals of Electronic Circuit Design*, John Wiley & Sons Inc., 2003.
- Fiore James M., *Operational Amplifiers & Linear Integrated Circuits: Theory and Application*, West Publishing Co., 1992.
- Floyd Thomas L. and Buchla David L., Electronics Fundamentals: Circuits, Devices and Applications, Pearson Education Limited, 2014.
- Franco S., *Design with Operational Amplifiers and Analog Integrated Circuits*, McGraw-Hill, 2001.
- Franco Sergio, *Design with Operational Amplifiers and Analog Integrated Circuits*, Mcgraw-Hill Education, 2015.
- Graeme Jerald G., Tobey Gene E., and Huelsman Lawrence P., *Operational Amplifies: Design and Applications*, Mcgraw-Hill, 1971.
- Gray P.R., Hurst P.J., Lewis S.H., and Meyer R.G., *Analysis and Design of Analog Integrated Circuits*, Wiley, 2001.
- Gray Paul E. and Searle Campbell L., *Electronic Principles: Physics, Models and Circuits*, John Wiley & Sons Inc., 1970.
- Guillermo Gonzalez, *Foundations of Oscillator Circuit Design*, Artech House Inc., 2007.
- Hayt W.H. and Neudeck G.W., *Electronic Circuit Analysis and Design*, Houghton Mifflin Co., 1984.
- Hiscocks Peter D., *Analog Circuit Design*, John Wiley & Sons Inc., 2006.
- Horenstein M.N., *Microelectronic Circuits and Devices*, Englewood Cliffs, Prentice-Hall, 1995.
- Horowitz Paul and Hill Winfield, *The Art of Electronics*, Cambridge University Press, 2015.
- Howe R.T. and Sodini C.G., *Microelectronics-An Integrated Approach*, Prentice-Hall, 1997.
- Jaeger R.C. and Blalock T.N., *Microelectronic Circuit Design*, McGraw-Hill, 2004.
- Johns D.A. and Martin K., *Analog Integrated Circuit Design*, Wiley, 1997.
- Michael Tooley, *Electronic Circuits: Fundamentals and Applications*, Newnes, 2002.
- Millman J. and Grabel A., *Microelectronics*, McGraw-Hill, 1987.

- Millman Jacob and Halkias Christos C., *Electronic Devices and Circuits,* McGraw-Hill Inc., 1967.
- Millman Jacob and Halkias Christos C., *Integrated Electronics: Analog and Digital Circuits and Systems,* McGraw-Hill International Student Edition, 1972.
- Muller Richard S. and Kamins Theodore I., *Device Electronics for Integrated Circuits,* Wiley, 2003.
- Neamen Donald A., *Electronic Circuit Analysis and Design,* McGraw-Hill, 2001.
- Neamen Donald A., *Microelectronics: Circuit Analysis and Design,* McGraw-Hill, 2010.
- Neamen Donald A., *Semiconductor Physics and Devices: Basic Principles,* McGraw-Hill, 2003.
- Neudeck George and Neudeck Gerold W., *The Bipolar Junction Transistor,* Addison-Wesley, 1989.
- Nihal Kularatna, *Electronic Circuit Design: From Concept to Implementation,* CRC Press, 2008.
- Pulfrey D.L. and Tarr N.G., *Introduction to Micro-electronic Devices,* Prentice-Hall, 1989.
- Rashid Muhammad H., *Microelectronic Circuits: Analysis and Design,* Cengage Learning, Inc., 2009.
- Razavi Behzad, *Design of Analog CMOS Integrated Circuits,* McGraw-Hill, 2001.
- Razavi Behzad, *Design of CMOS Analog Integrated Circuits,* Mcgraw-Hill, 2001.
- Razavi Behzad, *Fundamentals of Microelectronics,* John Wiley & Sons Inc., 2013.
- Roberge James K., *Operational Amplifiers: Theory and Practice,* John Wiley & Sons, 1975.
- Schilling D.L. and Belove C., *Electronic Circuits,* McGraw-Hill, 1979.
- Schultz Mitchel E., *Grob's Basic Electronics,* McGraw-Hill Education, 2016.
- Searle C.L., Boothroyd A.R., Angelo E.J., Jr., Gray P.E., and Pederson D.O., *Elementary Circuit Properties of Transistors, Vol. 3 of the SEEC Series,* Wiley, 1964.
- Sedra Adel S. and Kenneth C., *Microelectronic Circuits,* Oxford University Press, 2004.
- Sinclair Ian R. and Dunton John, *Practical Electronics Handbook,* Elsevier Ltd., 2007.
- Smith J.L., *Modern Operational Circuit Design,* Wiley Interscience, 1971.
- Sodagar Amir M., *Analysis of Bipolar and CMOS Amplifiers,* CRC Press, 2007.
- Spencer R.A. and Ghausi M.S., *Introduction to Electronic Circuit Design,* Pearson Education Inc., 2003.
- Stephan J. G. Gift and Brent Maundy, *Electronic Circuit Design and Application,* Springer, 2021.
- Streetman B.G. and Banerjee S., *Solid-State Electronic Devices,* Prentice-Hall, 2000.
- Terrell David L., *Op Amps: Design, Application, and Troubleshooting,* Elsevier Science, 1996.
- Tietze Ulrich, Schenk Christoph, and Gamm Eberhard, *Electronic Circuits: Handbook for Design and Application,* Springer, 2008.
- Tsividis Y., *Operation and Modeling of the MOS Transistor,* Oxford University Press, 1999.
- Van Valkenburg M.E., *Analog Filter Design,* Oxford University Press Inc., 1982.
- Wait J.V., Huelsman L.P., and Korn G.A., *Introduction to Operational Amplifiers Theory and Applications,* McGraw-Hill, 1975.
- Whitaker Jerry C., *The Electronics Handbook,* Taylor and Francis, 2005.